Competency Based Mathematics

for

Secondary Schools

Book 1

(MODULES 1 TO 4)

Nji Emmanuel Ndi

GBHS Mankon, Bamenda

North West Region, Cameroon

Tel: (+237) 676 684 050

Email: manuelndike@gmail.com

First Edition

Printed by CreateSpace, an Amazon.com Company

EStore address: www.CreateSpace.com/6812322

Available from Amazon.com, CreateSpace.com, and other retail outlets

Available on Kindle and other retail outlets

<u>Books by Nji Emmanuel Ndi:</u>

Complete Ordinary Level Mathematics Passport

Rudiments of Ordinary Level Mathematics

Advanced Level Pure Mathematics Key Facts

Competency Based Mathematics for Secondary Schools Book 1

Competency Based Mathematics for Secondary Schools Book 2

Competency Based Mathematics for Secondary Schools Book 3

DEDICATION

Dedicated to all emerging and
emergent Societies

Table of Contents

Acknowledgement

My deepest gratitude goes to God Almighty for the inspiration and for the strength.

Many thanks go to Mme. Mbuameh Daisy and Mr. Mburubah Walters for their critical proof reading of the typescript and for offering very useful suggestions which went a long way to reshape the work, the North West Regional Pedagogic Inspector for Mathematics Mr. Nfor Samuel Ndi who preview the initial manuscript and gave ample advice, which went a long way to reshape the document. I heartily thank the Former North West Regional Pedagogic Inspector for Mathematics Mr. Nji Samuel Tatah who made a very commendable effort to edit the Mathematics content of the book. I cannot forget the last minute encouragements and advice which the National inspector of Mathematics Mme Babila Emilia inspired me with. I equally pay much tribute to my students on which this material was tested. I cannot end here without thanking my sweet heart Nji Irene Nfih and my Children who encouraged and supported me in one way or the other during the course of the work.

Many thanks go to the WAEC and the CGCE Board for allowing their past questions to be used directly or indirectly.

Nji Emmanuel Ndi

G.B.H.S. Mankon, Bamenda

North West Region

Cameroon

TEL: (+237)76684050

E-mail: manuelndike@gmail.com

How to Use this Book

This book is written in a very special way with different sections boxed and represented by special symbols as follows.

| ? | Brainstorming Exercise |

| Example |

| Competency Based Exercise 2:1 |

| Exercise |

| Skill Building Exercise |

| Discussion Exercise |

| Integration Activity |

| Investigative Activity |

| Multiple Choice Exercise |

| Review Exercise |

| Group Activity |

The various sections represented by different symbols are out to facilitate navigation through the book. By investing enough time and energy in each section both students and teachers will realize that their speed and understanding will be greatly enhanced.

The brain storming exercises are aimed at provoking and invoking the learners' minds to prepare them for the task at hand. The teacher is highly encouraged to orally question the students during lessons using questions under this section.

The investigative exercises are meant to give the learner ample opportunity to experiment and self-discover facts and concepts and develop methods and skills without being told.

The group activities and discussion exercises are aimed at developing a team spirit in the learners.

Many well designed examples are vividly used and solved to facilitate the learner's understanding by showing the necessary steps required for a particular solution. There are a good number of real life examples which point out the application of the subject matter in real life situations. The student is advised to study these examples very carefully.

There are many well graded exercises and skill building exercises to test the level of understanding of the learner and to facilitate skill development in the learner. The student is advised to attempt all the questions as each question may have its own technique.

Many integration activities have been designed to unify groups of sub topics, topics or modules in some cases.

Where necessary review exercises have been given to help the learner retain the skills acquired in the earlier sections.

Finally each topic ends with a good number of multiple choice questions. In each question only one of the alternatives is correct. Write down the letter corresponding to the correct answer.

For greatest achievement, the learner is advised to study regularly what he does not know and work without fear of making mistakes whether with the teacher or during group work.

By consistently and systematically going through this course as instructed, the learner will be overwhelmed with the competencies acquired at each level and at the end of the course.

Notations Used in this Book

+	Addition
−	Subtraction
×	Multiplication
÷	Division
%	Percentage
$\mathbb{Z}$	The set of integers, $\{0, \pm 1, \pm 2, \pm 3, \pm 4, \ldots\}$
$\mathbb{N}$	The set of all positive integers and zero, $\{0, 1, 2, 3, 4, \ldots\}$
$\mathbb{Z}^+$	The set of positive integers $\{+1, +2, +3, +4 \ldots\}$
$\mathbb{Q}$	The set of rational numbers
$\mathbb{Q}^+$	The set of positive rational numbers
$\mathbb{R}$	The set of all real numbers $\{x: x \in \mathbb{R}\}$
$\mathbb{R}^+$	The set of all positive real numbers $\{x \in \mathbb{R}: x > 0\}$
=	Is equal to
≠	Is not equal to
≈	Is approximately equal to
<	Is less than
>	Is greater than
$\sqrt{}$	The positive square root
$\sqrt[3]{}$	Cube root
⊥	Is perpendicular to
‖	Is parallel to
∠	Angle
°	Degree
°C	Degrees Celsius
°F	Degrees Fahrenheit
$\sum$	Summation

Module 1

Numbers, Fundamental Operations and Relationships in the set of Numbers

Family of Situations

At the end of module 1, the student is expected to have acquired competencies within the **families of situations** *'Representation, determination of quantities and identification of objects by numbers'.*

Categories of Action

The categories of action for module 1 include:
1. Determination of a numbers,
2. Reading and writing information using numbers,
3. Verbal interaction on information containing numbers
4. Estimation and treatment of quantities.

Credit

The module is expected to be covered within 8 weeks teaching 4 hours per week (or within 30 to 32 hours).

Topic 1

NUMBERS AND NUMERALS

Objectives

At the end of this topic, the learner should be able to:

1. Distinguish between numbers and numerals.
2. Recognize numerals as symbols used in representing numbers.
3. Give a brief historical account of the development of number systems.
4. Represent numbers using Egyptian, Roman and Hindu-Arabic numerals.
5. Determine and state the value of a digit in a given Hindu-Arabic numeral.
6. Read and write in words a whole number given in symbols and vice versa.

1.1   Notion of Numbers and Numerals

A **number** is an idea that expresses a quantity, or that expresses how many things have been counted. Being an idea, a number can neither be seen nor touched. Thus, one can see or touch two pencils, but cannot touch the number two. Numbers are usually represented by symbols called **numerals**. For instance the number 'one', can be represented by the numerals 1, I, i etc; the number, 'two' can be represented by the numerals 2, II, ii etc.

1.2   Historical Development of Number Systems

The system of counting of the ancient civilizations consisted only of one, two, and many. These counting systems gradually grow until about 3000 BC when many civilizations began to write. At that time, they represented numbers by marking strokes on stones or clay or tying stones or sticks in bags.

The figure below shows how people of the early civilization represented three cattle.

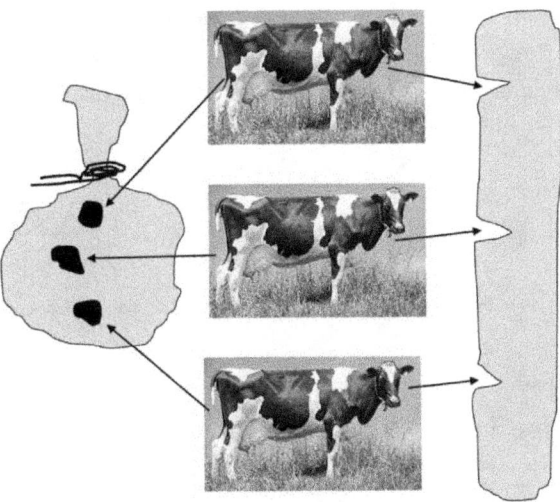

1.3   Egyptian, Roman and Hindu-Arabic Numerals

(a) The Egyptian Numerals

The following are the symbols which the Egyptian used to represent numbers.

Number	Numeral
One	I
Ten	∧
One hundred	9
One thousand	⚡
Ten thousand	ſ

In the Egyptian system, the symbols were written in any order.
Also in this system some large numbers are written using only very few symbols while some small numbers are written using so many symbols.

Example

Write the following using Egyptian numerals
(a) two hundred and forty nine   (b) ten thousand one hundred
(c) twenty three thousands four hundred and twenty one

Solution

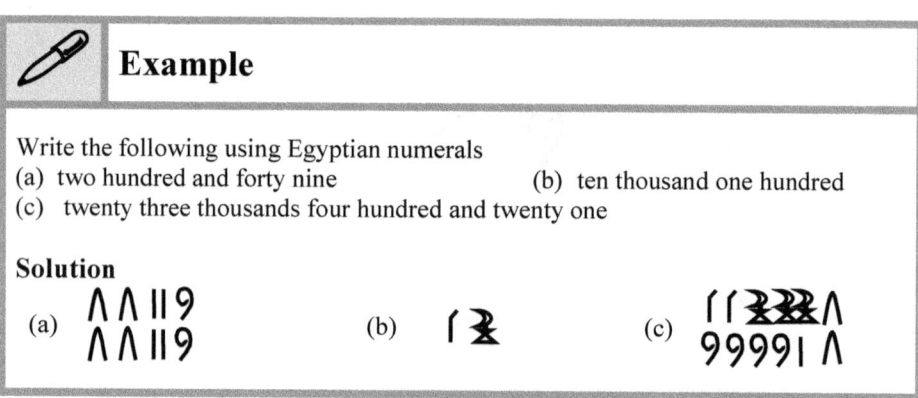

Exercise 1:1

1.   In what groups is the Egyptian system of numeration based?
2.   Write the following using Egyptian numerals.
   (a) 179   (b) 3478   (c) 6351   (d) 421   (e) 2456
   (f) 1334   (g) 18   (h) 1888   (i)   4478   (j) 286
3.   What do the following Egyptian numerals stand for?

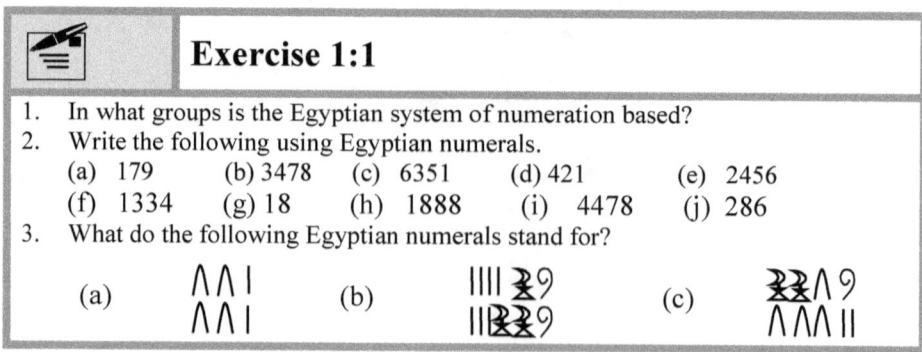

(d)  ||||99∧⅔
||||99∧⅔

(e)  ∧∧∧∧
∧∧⅔91

(f)  ∧∧9⅔|
∧99|||

(g)  9999∧||
999∧∧||

(h) 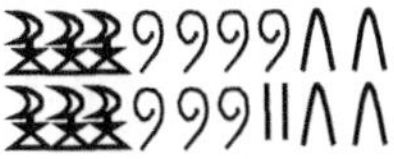 ⅔⅔⅔9999∧∧
⅔⅔⅔999||∧∧

(i) 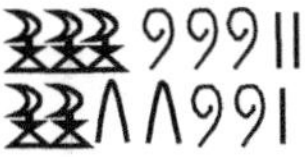 ⅔⅔⅔999||
⅔⅔∧∧991

(j) 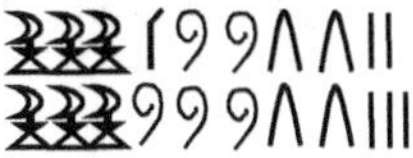 ⅔⅔⅔ſ99∧∧||
⅔⅔⅔999∧∧|||

4. Add the following like an Egyptian.

(a) 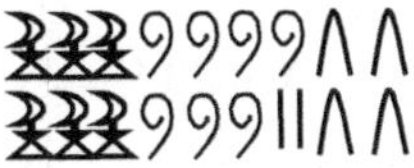 ⅔⅔⅔9999∧∧
⅔⅔⅔999||∧∧   and   ⅔⅔⅔999∧||
⅔⅔∧∧99∧|

(b)  ⅔⅔⅔99∧
⅔⅔⅔9∧∧| and   ⅔⅔99∧|
⅔⅔99∧|

5. An Egyptian newspaper reported that during a flood of the R. Nile, out of the

ſſ⅔⅔⅔∧||
9999∧| people along the R. Nile, many people died leaving only

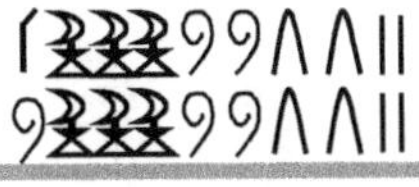

ſſ⅔⅔⅔99∧∧||
9⅔⅔⅔99∧∧|| . How many people died during the flood?

(b) The Roman Numerals
The table below shows the symbols used in the Roman number system.

Number	Numeral
One	I
Five	V
Ten	X
Fifty	L
One hundred	C
Five hundred	D
One thousand	M

5

In the Roman number system mostly capital letters of the English alphabet are used. Sometimes the   small letters i, v and x are used to represent one, five and ten. The above symbols are used in different combinations to represent numbers even more than a million. For instance

1998 is written in Roman numerals as MCMXCVIII

754 is written in Roman numerals as DCCLIV

Remarks

(a) For the Romans unlike the Egyptians, the order in which the symbols were written was very important. For instance, while the Egyptians could write Λ9 or 9Λ to represent one hundred and ten, for the Romans, CX, represented one hundred and ten and XC, represented ninety.

(b) The Roman system of numeration is built on "five", probably because they used the five fingers of one hand to count. Thus, they wrote V instead of IIIII, L instead of XXXXX, D instead of CCCCC and so on. Such a system whereby counting is done in groups of fives is called a **base five system** of numeration.

(c) A letter placed after another of greater value, adds to its value.

e.g.   VI = 5 + 1 = 6.

(d) On the other hand, a letter placed before another letter of greater value, subtracts from its value e.g. IV = 5 – 1 = 4.

Some Uses of Roman Numerals

Roman numerals unlike Egyptian numerals are used very often in our day to day life to:

    (i)   Number chapters, pages and exercises in textbooks.
    (ii)   Number some clock faces.
    (iii)   Number result sheets and articles of constitutions.
    (iv)   Matriculate cars.

Exercise 1:2

1. Write the following using Roman numerals
    (a) 29   (b) 48   (c) 5874   (d) 8993   (e) 1338   (f) 1990
    (g) 3000   (h) 6432   (i) 549   (j) 7428   (k) 9478   (l) 286
2. Write the following Roman numerals in words.

    (a) CLIV   (b) DIX   (c) MCXIV   (d) DCCLXI

    (e) DLXXXIX   (f) MCMXCIX   (g) LIX   (h) MCD

    (i) MMMCDXL   (j) DCCCI   (k) MMCDLC   (l) MMCD
3. List the Roman numbers in increasing order of magnitude from one to fifty.
4. What Roman numerals do the following Egyptian numerals represent?

    (a) 999∧
         99∧II

    (b) ₤9∧II
         ₤9∧I

    (c) ₤₤99IIII
         ₤₤999III

    (d) 99∧III
         9∧∧₤I

    (e) ∧∧II
         ∧∧ II

    (f) 99IIII
         9∧∧I

5. What Egyptian Numerals stand for the following Roman numerals?
    (a) MDCCLVIII   (b) LXXXVII   (c) XLV

    (d) DCLXV   (e) MMCCLXXI   (f) MCMXCVIII

(c) The Hindu-Arabic Numerals

Our number system is called the Hindu- Arabic system. Originally, these numerals were so different from what exist today. Even today, the Arabic numerals used in the Islamic world are very different from the ones used internationally. Compare the two in the table below.

Hindu-Arabic	1	2	3	4	5	6	7	8	9	0
Islamic world	١	٢	٣	٤	٥	٦	٧	٨	٩	٠

1.4 The Place Value System

In the Hindu-Arabic number system, the order of the symbols is very import.

Each of the symbols 0, 1, 2, 3, 4, 5, 6, 7, 8 and 9 is called a **digit** and the value of each symbol depends on its position or place in a numeral. For instance, the value of 3 in 846302 is three hundred and in 8463002 is three thousands. For this reason the Hindu-Arabic system is a place -value system.

The Hindu-Arabic system is also built on 'ten'. A system of numeration in which counting is done in groups of ten is called a base ten, denary or decimal system.

Therefore, the Egyptian system is a base ten system, but not a place value system.

It should be noted that, a place-value system can use any base, one of the symbols must be zero and the number of digits must be equal to the base used.

The place value system makes it possible for numbers to be written without any confusion. Thus bills such as water and electric bills are easily written and read by the public. Without the place value system, it will be very difficult to distinguish say a bill of four thousands and fifteen (4015) francs from a bill of four hundred and fifteen (415) francs.

 Exercise 1:3

1. State the value of 3 in each of the following.
   (a) 37570   (b) 613004   (c) 9320161
2. Explain why 68 is not the same as 86?
3. Write in words (a) 5578   (b) 50448   (c) 893261   (d) 17204
4. Write each of the following using Modern Hindu-Arabic numerals
   (a) Five hundred and thirty eight thousand and one.
   (b) Seventeen thousand and four
   (c) Nine thousand nine hundred and nine
   (d) Two hundred and thirty two
   (e) One hundred and eleven thousand one hundred and one
   (f) Eight thousand and Eighty
   (g) (g) Ten thousand and ten.

 Integration Activity 1

1. You are a CRTV journalist on special mission in Rome and you have to send a report to the CRTV English news desk concerning an election that have just taken place in Rome. A Roman journalist gives you his data sheet with the following results.

S/N	Name of Party	Number of Votes
1	Liberal party	MCDXLVIII
2	Communist party	MMCDXLIX
3	Socialist party	MCMXCIX
4	Null ballots	MDXLIV

(a) Complete the following written report.

According to a report just filed to our news desk, _____ people registered for the elections and there were _____ null ballots. Therefore _____ people actually voted. The _____ Party is leading with _____ votes, closely followed by the _____ Party which has _____ votes. The _____ Party is the third on the list with _____ votes.

(b) An Egyptian news agency pays you to rewrite your report so that the numbers will be in Egyptian numerals. Complete the report in (a) using Egyptian numerals as required.

(c) You are called to relay your report in (a) to CRTV. Read your report to the hearing of your mates and teacher as you will relay it.

2. Your mother is a potato farmer. In her absence, a potato dealer calls you to inquire the price of a full load of potatoes which you were told is 1,090,205 francs.

(i) Tell him the price of the load. If the cost of transport is 80405 francs, inform him about this cost and the total cost when the potatoes arrive Douala.

(ii) The potato dealer pays for the load. Complete the following receipt which you will give to him against this payment.

Self-Reliance Potato Farmer

Receipt No: 67419

Date ....

Received from *Mr. Yongabi Humphrey Sama* ....

The sum of ....

(In words)

....

Being *the cost of one load of potatoes* ....

In figures | 1,090,205 | FCFA.

Customer's Sign: *Yongabi Humphrey*   Receiver's Sign: ....

---

3. Your father has given you the following cheque to fill it and use it to withdraw 347165 francs from his account for your school fee, books, uniforms and other school needs. Fill the cheque as you would to withdraw the money.

Cheque No: 632261   UNITY BANK

Pay against this cheque the sum of _____XAF [ ]

_____

(In words)

To the order of_____ Signature_____

Issued at_____ on _____

Account Number 10029 26020 01113690701 23

Wanto Julius Fai   Signature *Wanto julius*

✎ Multiple Choice Exercise 1

1. The statement that refers to numerals is:
    [A] When counting 7 comes before 8   [B] The sum of 2 and 6 is 8
    [C] In 78, 7 comes before 8   [D] 78 is the sum of 50 and 28
2. The statement that refers to numbers is:
    [A] In 23, 2 comes before 3
    [B] When counting, 2 comes before 3
    [C] 2 combined with 3 is either 23 or 32
    [D] 23 consist of 2 and 3
3. The Hindu Arabic numeral representing "Two hundred and four thousand and four" is:
    [A] 20404   [B] 240004   [C] 24400   [D] 204004
4. The number, which represents one million three hundred and fifty four is:
    [A] 1000354   [B] 1030054   [C] 1300054   [D] 1354000
5. Ninety nine thousand and ninety nine written in figures is:
    [A] 990099   [B] 9999   [C] 99099   [D] 90999
6. We can read the number 605, 080 as:
    [A] Sixty thousand and five thousand and eighty
    [B] Six hundred and five hundred and eighty
    [C] Six thousand and five hundred and eighty

[D] Six hundred and five thousand and eighty

7. We can read the amount 2,300,240 francs as:
    [A] Twenty three million four hundred and twenty francs
    [B] Two million, three hundred thousand four hundred and twenty francs
    [C] Two million three hundred thousand two hundred and forty francs.
    [D] Twenty three million two hundred and forty francs

8. The value of the digit 6 in the number 726251 is:
    [A]   six hundred   [B]   six hundredth
    [C]   six thousandth   [D]   six thousand

9. The value of 5 in 2753 is:
    [A] Tenth   [B] tens   [C] Hundredth   [D] hundredth

10. When we divide the value of 6 in 5624 by the value of 3 in 2639, the result is:
    [A]   2   [B]   5   [C]   20   [D]   16

11. The product of the value of 7 in 2721 and the value of 3 in 5837 is:
    [A]   21000   [B] 26677   [C]   21   [D]   2100

12. The sum of eleven thousand and one thousand hundred is:
    [A]   11100   [B]   12100   [C]   11110   [D]   111000

13. In 6,367,804, the value of the underlined digit is:
    [A] 7   [B]   700   [C] 7,000   [D] 70,000

14. We can write four million and six as:
    [A] 4,000,600   [B]   4,000,006   [C]   4,600   [D]   4,006

15. The Roman numeral **CDXXXVII** represents:
    [A] 437   [B] 637   [C] 187   [D] 487

16. The Roman numeral **CMXXII** represents:
    [A] 5220   [B] 1922   [C]   922   [D] 918

17. The number 40 in Roman numerals is:
    [A] XV   [B] XL   [C] XLX   [D] LX

18. 1167 in Roman numerals is:
    [A] MCLXII   [B] DCLXVII   [C] MCLXVII   [D] MCXLVII

19. As a Roman numeral 2598 is the same as:
    [A] IIMVCIXVIII   [B] MMDXCVIII   [C] MMIIDC   [D] XXVIIC

20. In Roman numeral 548 is:
    [A] DXLVIII   [B] IIXLD   [C] DLIIX   [D] VIIILD

21. The number 1200 in Egyptian numerals is:

    [A]   ˙ΛΛ   [B] 9̇ ˙ ˙   [C]   9̇ 9̇   [D]   99

22. The Egyptian numeral 9̄ΙΛΙ stands for the Hindu-Arabic numeral:
    [A] 91200   [B] 991   [C] 11111   [D] 1112

23. The Hindu Arabic numeral representing "Two hundred and four thousand and four" is:
    [A]   20404   [B] 240004   [C]   244004   [D] 204004

11

Topic 2

NATURAL NUMBERS

Objectives

At the end of this topic, the learner should be able to:

1. Recognize ℕ (not N) as the set of natural numbers.
2. Distinguish between counting numbers and natural numbers.
3. Compare natural numbers using the symbols < and >.
4. Use natural numbers in different circumstances such as counting, identification etc.
5. Add and multiply natural numbers.
6. State and use the properties of addition and multiplication to manipulate natural numbers.
7. Perform mixed operations involving addition and multiplication.
8. Write powers of natural numbers in index form and vice versa.
9. Count and write consecutive numbers in base ten and other bases.
10. Write numbers in expanded and condensed form.
11. Convert from a given base to base ten and vice versa.
12. Convert from a non-denary base to another non-denary base.

2.1 Counting Numbers and Natural Numbers

?	**Brainstorming Exercise**

1. When counting, what numbers do we use?
2. Which number do we begin to count from?
3. What is the general name of the numbers used in counting?

When counting, objects such as the number of students in a class, we use the whole numbers beginning from one. Thus the set of **counting numbers** is denoted by $\mathbb{N}^*$ where $\mathbb{N}^* = \{1, 2, 3, 4, 5...\}$. If we recognize that there can be no student in the class then it makes sense for us to include 0 to this group of numbers. Therefore we can extend the set of counting numbers to the set $\{0,1,2,3 ...\}$ called the set of **Natural numbers** usually denoted by $\mathbb{N}$ (not **N**).

Thus $\mathbb{N} = \{0, 1, 2, 3, 4, 5 ...\}$ and $\mathbb{N}^* = \{1, 2, 3, 4, 5 ...\}$.

2.2 Comparing Natural Numbers (Ordering in $\mathbb{N}$)

?	**Brainstorming Exercise**

1. Which of 2 and 5 is greater?
2. Which of 2 and 5 is smaller?
3. What symbols can we use to express the ideas in 1 and 2?

We use the symbol $=$ read "is equal to" to show that two things or numbers are equal. For instance, 7 days $=$ 1 week.

No two natural numbers are equal. When two things are not equal, the symbol $\neq$ is used. However to be precise, we use the symbol

(i) $<$ which means 'is less than' to express the idea that one thing or number is less than another. For instance $2< 7$ is read '2 is less than 7'.

(ii) $>$ which means 'is greater than' to express the idea that one thing or number is greater than another. For instance $10 > 7$ is read '10 is greater than 7'.

Notice that the symbols are read from left to right.

Using these symbols, we can write the following equivalent relationship between the first ten natural numbers.

$$0 < 1 < 2 < 3 < 4 < 5 < 6 < 7 < 8 < 9$$

or

$$9 > 8 > 7 > 6 > 5 > 4 > 3 > 2 > 1 > 0$$

Competency Based Exercise 2:1

1. The following shows the number of votes obtained by the five candidates who contested for the post of Senior Prefect in a certain school.

Candidate Name	Number of votes
Suh	30
Ngwa	16
Anye	32
Tama	14
Chi	22

Classify the candidates using
(a)   The symbol >.   (b)   The symbol <.
(c)   Ordinal numbers (i.e. first, second, etc)

2. The student union of your village decides to give scholarships to some members using the following criteria.
Anye should have 2000 francs more than Doh.
Bih should have 6000 francs.
Chi should have 2000 francs less than Anye.
Doh should have 4000 francs more than Bih.
Eba should have 3000 francs more than Anye.

You are the financial secretary of the union. Classify the students according to the amounts they should receive from least to the highest stating the amount received by each student against his/her name.

 Skill Building Exercise 2:1

(1) Arrange 75077, 98706, 60793 and 30676 in order of magnitude using
 (a) the symbols <   (b) the symbols >
(2) Use the symbols < or > to compare the following
 (a) 3....7   (b) 10....8   (c) 14....13   (d) 4...6
 (e) 5−1....9   (f) 26−3....5   (g) 6−3....2   (h) 4+2....8
 (i) 2...7−4   (j) 11...5+8   (k) 3 × 4 2+7   (l) 18÷6 ... 7
 (m) 458   764   (n) 963   687   (o) 8463 8379   (p) 6536 8543
(3) Using the symbol < arrange the numbers 56, 74, 121, 68.
(4) Using the symbol > arrange to the numbers 585, 847, 497, 901.

14

Some Uses of Natural Numbers

Natural numbers are very useful in the following ways:

(i)   As cardinal numbers in counting objects.
(ii) As nominal numbers to classify objects in a given order such as from first to last.
(iii) In calculations to add, subtract, multiply, divide etc.
(iv) For identification such as in car number plates, identity card numbers, secret codes, candidate examination numbers etc.

2.3 Addition and Multiplication of Natural Numbers

? Brainstorming Exercise

1. In $2 + 5 = 7$, what is the name given to the numbers 2 and 5?
2. What is the name given to the result obtained by adding numbers?
3. In $2 \times 5 = 10$, what is the name given to the number 2?
4. In $2 \times 5 = 10$, what is the name given to the number 5?
5. What is the name given to the result obtained by multiplying numbers?

Addition ($+$), is also referred to as the **sum or plus**. The numbers to be added are called **addends.** The result of the addition is also called the **sum.**

$$\overset{\text{addend}}{5} \;+\; \overset{\text{addend}}{7} \;=\; \overset{\text{sum}}{12}$$

Multiplication ($\times$), is also called **times** or **product**. In multiplication, the number being multiplied is called the **multiplicand** and the number by which the **multiplication** is to be performed is called the **multiplier**. The result of the multiplication is also called the **product.**

$$\overset{\text{multiplicand}}{4} \;\times\; \overset{\text{multiplier}}{6} \;=\; \overset{\text{product}}{24}$$

Properties of Addition and Multiplication

 Investigative Activity

1. Evaluate the following.
   (a) $2 + 5$ (b) $5 + 2$ (c) 6×3 (d) 3×6
   (e) $(2+5)+3$ (f) $2+(5+3)$ (g) $(6 \times 3) \times 2$ (h) $3 \times (6 \times 2)$
   (i) $0+3$ (j) $3 + 0$ (k) 0×3 (l) 3×0
   (m) $3 \times (2+5)$ (n) $3 \times 2 + 3 \times 5$ (o) $4 \times (6 + 3)$ (p) $4 \times 6 + 4 \times 3)$
   (k) 1×3 (l) 3×1

2. What conclusion do you draw about the result of the addition of two natural numbers and the result when the two numbers are interchanged before adding them?

3. Does your conclusion in (2) above hold for multiplication?

4. What conclusion do you draw about the result of a sum of more than two natural numbers when the numbers are grouped differently?

5. What conclusion do you draw about the result of adding 0 to a number or adding a number to 0?

6. What conclusion do you draw about the result of multiplying 0 by a number or multiplying a number to 0?

7. What conclusion do you draw about the result of the product of a number and a sum of two numbers and the sum of the separate products of a number and two numbers?

8. Does your conclusion in (4) above hold for multiplication?

9. What conclusion do you draw about the result of multiplying 1 by a number or multiplying a number to 1?

10. Are the results in each case also natural numbers?

1. The sum (or product) of two natural numbers is the same when the numbers are interchanged.

   Examples
   (a) $2 + 5 = 5 + 2$ (b) $3 \times 4 = 4 \times 3$

2. The sum (or product) of more than two natural numbers is the same no matter the way the numbers are grouped.

   Examples
   (a) $(2 + 5) + 8 = 7 + 8 = 15$ and $2 + (5 + 8) = 2 + 13 = 15$
   (b) $(3 \times 4) \times 2 = 12 \times 2 = 24$ and $3 \times (4 \times 2) = 3 \times 8 = 24$

3. The sum of zero and any natural number is the number.
   The product of zero and any natural number is zero.

   Examples
   (a) $0 + 3 = 3$ and $3 + 0 = 3$ (b) $0 \times 5 = 0$ and $5 \times 0 = 0.$

4. When we multiply any number by 1 or multiply 1 by any number the result is the number.

   Examples
   (a) $1 \times 3 = 3$ and $3 \times 1 = 3$ (b) $1 \times 7 = 7$ and $7 \times 1 = 7$

5. The product of a number and a sum of numbers is equal to the sum of the separate products of the number and the numbers.

   Examples
   (a) $3 \times (2 + 5) = 3 \times 2 + 3 \times 5 = 6 + 15 = 21$
       $3 \times (2 + 5) = 3 \times 7 = 21$
   (b) $4 \times (6 + 3) = 4 \times 6 + 4 \times 3 = 24 + 12 = 36$
       $4 \times (6 + 3) = 4 \times 6 + 4 \times 3 = 24 + 12 = 36$

6. The sum (or product) of two natural numbers is also a natural number.

   Examples
   (a) $5 \in \mathbb{N}$, $3 \in \mathbb{N}$ and $5 + 3 = 8 \in \mathbb{N}$
   (b) $7 \in \mathbb{N}$, $4 \in \mathbb{N}$ and $7 + 4 = 11 \in \mathbb{N}$

Applications of the Properties of Numbers for Addition

We call any two digits whose sum is 10 **complementary digits**. For instance 4 and 6 are complementary digits and 4 is the complement of 6 because $4 + 6 = 10$.

 Discussion Exercise

1. How can you use complementary digits and the associative property to find the sum of two digits whose sum is greater than 10?
2. Use the method you have discussed in (1) above to mentally and speedily evaluate   (a) $8 + 7$ (b) $7 + 9$ (c) $9 + 9$

To add two digits whose sum is greater than 10, subtract the complement of the larger digit from the smaller digit and add to the larger to make 10 and then add the residue of the smaller digit to the 10. Thus,

$8 + 7 = 8 + (2 + 5) = (8 + 2) + 5 = 10 + 5 = 15$

$7 + 9 = (6 + 1) + 9 = 6 + (1 + 9) = 6 + 10 = 16$

Small numbers can easily be added mentally but larger numbers need to be arranged in columns according to the place value of the digits in such a way that units are in one column, tens are in one column, hundreds are in one column and so on.

Example

Evaluate: (i) 6435 + 3268   (ii) 8535 + 258

Solution

(i)   6435
     + 3264
     ‾‾‾‾‾‾
       9699

(ii)   85$\overset{1}{3}$5
          +   258
          ‾‾‾‾‾‾
            8793

 Skill Building Exercise 2:2

1. Evaluate the following without using calculators
    (a) 4365+85985   (b) 67454+43598   (c) 5324+541683+6745
2. The sum of 5 addends is 258. 4 is added to each of the addends. What is the new sum?

 Competency Based Exercise 2:2

You are in a partnership with four other people. The total shares amount to 426000 francs. The company decides to increase its shares by 28000 francs. What will be the new total shares?

Multiplication Skills

For small numbers multiplication can be done using multiplication tables like that below until the learner gains facility in multiplication to do it off hand.

×	0	1	2	3	4	5	6	7	8	9
0	0	0	0	0	0	0	0	0	0	0
1	0	1	2	3	4	5	6	7	8	9
2	0	2	4	6	8	10	12	14	16	18
3	0	3	6	9	12	15	18	21	24	27
4	0	4	8	12	16	20	24	28	32	36
5	0	5	10	15	20	25	30	35	40	45
6	0	6	12	18	24	30	36	42	48	54
7	0	7	14	21	28	35	42	49	56	63
8	0	8	16	24	32	40	48	56	64	72
9	0	9	18	27	36	45	54	63	72	81

On the multiplication table 7 × 6 has been done.
Verify that 9 × 7 = 63 and 6 × 8 = 48.

To multiply larger numbers, long multiplication can be used.

 Example

Evaluate (i) 453 × 7 (ii) 247 × 127

Solution

(i)

```
      4   5   3
  ×           7
  ─────────────
  3   1   7   1
```

Explanation

7 × 3 = 21. Write 1 and carry 2.
7 × 5 = 35. Plus 2, equal 37. Write 7 and carry 3.
7 × 4 = 28, plus 3 equal 31.

(ii) To do this, we can arrange the numbers as follows.

```
          2   4   7
  ×       1   2   7
  ──────────────────
      1   7   2   9
      4   9   4   0
  2   4   7   0   0
  ──────────────────
  3   1   3   6   9
```

Explanation

247 × 7 = 1729. Write 1729 with 9 under the unit digit column.
247 × 20 = 4940. Write 4940 with 0 under the unit digit column.
247 × 100 = 24700. Write 24700 with the last 0 under the unit digit column.

Add the three products.

 Competency Based Exercise 2:3

Your company in Competency Based Exercise 2:2 instead decide to increase the shares of each partner by 6000 francs. What will be the new total shares?

2.4 Order of Operations for Addition and Multiplication

? **Brainstorming Exercise**

In 4 × 5 + 2, state the result when we
1. We multiply 4 by 5 before adding 2.
2. We add 2 to 5 before multiplying by 4.
3. Which of them then is correct and why?

From the above it is clear that there is need for an established rule which determines which of addition or multiplication should be performed first in an expression involving the two operations.

The convention adopted by mathematicians is that:

In an expression, involving multiplication and addition, multiply before adding. The mnemonic 'MA' may help you remember this.

 Example

Evaluate the following
(i) $8 \times 3 + 6$　　(ii) $9 + 5 \times 7$　(iii) $5 + 4 \times 6 + 9$　(iv) $2 \times 12 + 20 \times 5$

Solution
(i)　$8 \times 3 + 6 = 24 + 6 = 30$　　　　　(ii)　$9 + 5 \times 7 = 9 + 35 = 44$
(iii)　$5 + 4 \times 6 + 9 = 5 + 24 + 9 = 38$　(iv) $2 \times 12 + 20 \times 5 = 24 + 100 = 124$

 Skill Building Exercise 2:3

Evaluate the following.
(1) $15 \times 7 + 5$　　　　(2) $6 + 4 \times 6$　　　　(3) $7 + 6 \times 15$
(4) $9 \times 14 + 5$　　　　(5) $4 + 3 \times 8$　　　　(6) $3 \times 24 + 4$
(7) $5 \times 6 + 2 \times 5$　　(8) $7 \times 9 + 4 \times 10$　(9) $12 + 8 \times 6 + 64$
(10) $5 + 6 \times 4 + 20$　(11) $3 + 8 \times 2 \times 5$　(12) $6 \times 9 + 12 + 5$

2.5　Powers of Natural Numbers

? **Brainstorming Exercise**

1.　What is an easier way of writing $3 \times 3 \times 3 \times 3 \times 3 \times 3$?
2.　What is the meaning of 5^3?
3.　In 5^3, how do we call the 5 and how do we call the 3?

An easier way of writing $5 \times 5 \times 5 \times 5$ is 5^4 which means 'multiply 5 by itself four times'.

A number written as 5^4 is said to be in **index form** or in **exponential notation**. On the other hand $5 \times 5 \times 5 \times 5$ is said to be written in **expanded form**.

Notice that $5 \times 5 \times 5 \times 5 = 625$. A number such as 625 is said to be written in the **normal form** or **condensed form**.

 Investigative Activity

Use a calculator to compute the following.
(a) 5^1 (b) 9^1 (c) 17^1 (a) 5^0 (b) 9^0 (c) 0^0
What conclusion do you draw about raising a number to the power (i) 1?   (ii) 0?

➢ *Any number raised to the power 1 is the number.*
➢ *Any number other than 0 raised to the power 0 is 1.*
➢ 0^0 *is meaningless.*

 Skill Building Exercise 2:4

1. Write out the following in index form.
   (a) $2 \times 2 \times 2 \times 2 \times 2 \times 2 \times 2$ (b) $7 \times 7 \times 7 \times 7 \times 7$
   (c) $3 \times 3 \times 3 \times 3 \times 3 \times 3 \times 3 \times 3 \times 3$ (d) $6 \times 6 \times 6 \times 6 \times 6$
   (e) $8 \times 8 \times 8 \times 8 \times 8 \times 8 \times 8 \times 8$
2. Write the following in expanded form.
   (a) 9^2 (b) 5^3 (c) 7^6 (d) 2^8 (e) 3^4 (f) 10^2 (g) 10^5 (h) 10^4
3. Write the following in normal form.
   (a) 9^2 (b) 5^3 (c) 7^6 (d) 2^8 (e) 3^4 (f) 10^2 (g) 10^5 (h) 10^4
4. Write in index form.
   (a) 100,000   (b) 10,000,000   (c) 100,000,000

Powers of 10 are very useful in many instances and their manipulation should be taken very seriously.

NUMBER BASES

2.6 Vocabulary Associated with Number bases

Recall that the Hindu-Arabic number system is a **base ten (denary** or **decimal) system** and that the digits used in this system are 0, 1, 2, 3, 4, 5, 6, 7, 8, 9.

In any number base, the highest digit is one less than the number base and the number of digits is equal to the base.

2.7 The Concept of Number Bases

In base ten, 357 means 3 hundred + 5 tens + 7 units.

A businessman can count the number of pencils in a cartoon by arranging them as follows.

In so doing he can say he has 34 pencils left. This means that there are 3 groups of tens and 4 units left.

? Brainstorming Exercise

1. Another businessman instead counts in groups of eight. How many groups and how many ones will he have?
2. Represent your answer using digits.
3. How will you read your answer?

Clearly, there are 4 groups of eights and 2 ones left and we can represent this by 42_{eight} and read it as 'four two base eight'.

Brainstorming Exercise

1. Now arrange the pencils in groups of six. How many groups and how many ones do you have?
2. Represent your answer using digits.
3. Read your answer.

If your arrangement is correct, you should have 5 groups of six and 4 ones which we can write as 54_{six} and read it as 'five four base six.

Discussion Exercise

Discuss the steps you would take to express the base ten number, 27 as a number in base two.

We can group 27 units in twos as follows.

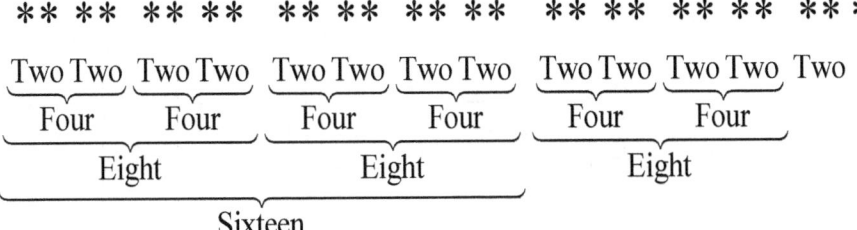

Therefore,
Twenty seven in base two
$$= 1 \text{ group of } 16 + 1 \text{ group of } 8 + 0 \text{ groups of } 4 + 1 \text{ group of } 2 + 1 \text{ unit}$$
$$\qquad 1 \qquad\qquad 1 \qquad\qquad 0 \qquad\qquad 1 \qquad\qquad 1$$
Twenty seven in base two $= 11011_{two}$.

We can also present this as follows.

27 units	= 13 groups of two and <u>1</u> unit left.
13 groups of two	= 6 groups of four and <u>1</u> group of two left.
6 groups of four	= 3 groups of eight and 0 groups of four left.
3 groups of eight	= 1 group of sixteen and <u>1</u> group of eight left.
1 group of sixteen	= 0 group of thirty two and <u>1</u> group of sixteen left.

To show the base to which a numeral is written, write the subscript spelling of the base after the numeral. For instance, if 1652 is in base seven we write 1652_{seven}, read ''one-six-five-two base seven' and **not** 'one thousand six hundred and fifty two base seven'. Note also that it is wrong to write 1652_7, because the

23

numeral 7 does not exist in base seven.

In base ten the base subscript is usually omitted. Thus, two hundred and seventy eight in base ten is written as 278_{ten} or simply 278.

Never omit the subscript notation for any numeral written in a base other than ten.

2.8 Counting in Different Bases

We can count in many different number bases. The following shows the first eleven consecutive numerals in the bases indicated.

Base 10	Base 9	Base 8	Base 7	Base 6	Base 5	Base 4	Base 3	Base 2
0	0	0	0	0	0	0	0	0
1	1	1	1	1	1	1	1	1
2	2	2	2	2	2	2	2	10
3	3	3	3	3	3	3	10	11
4	4	4	4	4	4	10	11	100
5	5	5	5	5	10	11	12	101
6	6	6	6	10	11	12	20	110
7	7	7	10	11	12	13	21	111
8	8	10	11	12	13	20	22	1000
9	10	11	12	13	14	21	200	1001
10	11	12	13	14	20	22	201	1010

 Exercise 2:1

1. Write 'Yes' if it is possible to write the given number in the base indicated, or 'No', if it is not possible. In each case give reasons for your answer.
   (a) 324_{six} (b) 471_{seven} (c) 1011_{two} (d) 897_{nine}
   (e) 952_{ten} (f) 782_{nine} (g) 342_{five} (h) 1846_{eight}
   (i) 3420_{five} (j) 2102_{three} (k) 764_{seven} (l) 1576_{eight}
2. Write the numbers from one to twenty in your native language.
3. What is the base used in counting in your native language?
4. Write down the first ten consecutive numerals in each of the following bases
   (a) base two   (b) base three   (c) base four
   (d) base five   (e) base seven   (f) base nine
5. Extend the table in the previous section 2.7.3 up to and including the thirtieth numeral in each number base.

2.9 Expanded Forms in Base Ten

In topic 1 it was stated that the value of each digit in a numeral depends on its position or place in the numeral. Hence, the numeral 5362 in base ten is interpreted as

$$5362 = 5 \times 1000 + 3 \times 100 + 6 \times 10 + 2$$
$$= 5 \times 10^3 + 3 \times 10^2 + 60 \times 10^1 + 2$$

The right hand side is called the **expanded numeral (form)** of 5362 while 5362 is called the **standard numeral** or **condensed form** of

$$5 \times 10^3 + 3 \times 10^2 + 60 \times 10^1 + 2$$

 Example

1. Write in expanded form. (i) 268   (ii) 42384   (iii) 735

   Solution
   (i) $268 = 200 + 60 + 8 = 2 \times 100 + 6 \times 10 + 8 = 10^2 + 6 \times 10 + 8$
   We usually write expanded forms directly without passing through the first two steps as above.

   (ii) $42384 = 4 \times 10^4 + 2 \times 10^3 + 3 \times 10^2 + 8 \times 10^1 + 4$
   (iii) $735 = 7 \times 10^2 + 3 \times 10 + 5$

2. Write the condensed form of   (a) $2 \times 10^3 + 5 \times 10^2 + 7 \times 10^1 + 9$
   (b) $4 \times 10^3 + 7 \times 10^2 + 9$

   Solution
   (a) $2 \times 10^3 + 5 \times 10^2 + 7 \times 10^1 + 9 = 2579_{ten}$
   (b) $4 \times 10^3 + 7 \times 10^2 + 9 = 4079_{ten}$

 Skill Building Exercise 2:5

1. Write down the expanded forms of each of the following base ten numerals
   (a) 87   (b) 124   (c) 3204   (d) 6007   (e) 96800   (f) 7050300
2. Write down the condensed forms for each of the following.
   (a) $4 \times 10 + 6$ (b) $7 \times 10^2 + 5 \times 10 + 2$ (c) $8 \times 10^3 + 3 \times 10^2 + 7$
   (d) $3 \times 10^3 + 5$ (e) $1 \times 10^4 + 6 \times 10^3 + 3 \times 10^2$ (f) $4 \times 10^6 + 6 \times 10^4 + 2 \times 10^2$

2.10 Expanded Forms in other Bases

Expanded forms can also be written in bases other than ten.

 Example

1. Write the expanded forms of   (a) 324_{five} (b) 2517_{eight} (c) 4702_{nine}

   Solution

   (a) $324_{\text{five}} = 3 \times 10^2 + 2 \times 5 + 4$ (b) $2517_{\text{eight}} = 2 \times 8^3 + 5 \times 8^2 + 1 \times 8 + 7$
   (c) $4702_{\text{nine}} = 4 \times 9^3 + 7 \times 9^2 + 0 \times 9 + 2 = 4 \times 9^3 + 7 \times 9^2 + 2$

2. Write the condensed form of
   (a) $3 \times 6^3 + 2 \times 6^2 + 5 \times 6 + 1$ (b) $6 \times 8^3 + 3 \times 8 + 7$

   Solution

   (a) $3 \times 6^3 + 2 \times 6^2 + 5 \times 6 + 1 = 3251_{\text{six}}$ (b) $6 \times 8^3 + 3 \times 8 + 7 = 6037_{\text{eight}}$

 Skill Building Exercise 2:6

1. Write the expanded forms of the following.
   (i) 3214_{five} (ii) 1011101_{two} (iii) 52301_{seven} (iv) 43021_{six}
   (v) 50032_{eight} (vi) 24002_{seven} (vii) 20121_{three} (viii) 34021_{eight}
   (ix) 7480_{nine} (x) 1101_{two} (xi) 4836_{nine} (xii) 2301_{four}
   (xiii) 400_{six} (xiv) 7104_{eight} (xv) 10221_{four} (xvi) 2100_{three}

2. Write down the condensed forms for each of the following.
   (a) $3 \times 4 + 2$ (b) $5 \times 8^2 + 7 \times 8 + 1$ (c) $2 \times 5^3 + 1 \times 5^2 + 4$
   (d) $6 \times 7^3 + 3$ (e) $4 \times 6^4 + 3 \times 6^3 + 1 \times 6^2$ (f) $3 \times 9^6 + 6 \times 9^4 + 8 \times 9^2$

2.11 Conversion of Number Bases

(i) Converting None Denary to Denary

To convert a numeral written in some base other than ten to base ten, the
numeral is written in expanded form and then evaluated completely.

 Example

Example 2:6
Convert the following to base ten   (i) 123_{six} (ii) 5724_{eight}

Solution
(i) $123_{\text{six}} = 1 \times 6^2 + 2 \times 6 + 3 = 36 + 12 + 3 = 51_{\text{ten}}$
(ii) $5724_{\text{eight}} = 5 \times 8^3 + 7 \times 8^2 + 2 \times 8 + 4 = 5 \times 512 + 7 \times 64 + 16 + 4 = 5724_{\text{ten}}$

Skill Building Exercise 2:7

Convert the following to base ten numerals

(1) 142_{five}	(2) 214_{five}	(3) 435_{six}	(4) 2542_{seven}
(5) 1126_{seven}	(6) 1011101_{two}	(7) 3247_{eight}	(8) 45631_{seven}
(9) 11011_{two}	(10) 3487_{nine}	(11) 11201_{three}	(12) 2103_{four}
(13) 35201_{six}	(14) 1824_{nine}	(15) 2112101_{three}	(16) 587234_{nine}

(ii) Converting Denary to None Denary

$20 \div 6 = 3$ remainder 2 can also be written as $20 = 6 \times 3 + 2$ (Since $6 \times 3 = 18$).

20, is called the **dividend**; 6, is called the **divisor**; 3, is called the **quotient** and 2 is called the **remainder**. This idea shall be employed to convert base ten numerals to non-base ten numerals as follows.

To change a base ten number to a number in a different base, repeated division by the destination base is done. In each step the remainder is written until the dividend is zero. The result is made up of the digits of the remainders taken in order from the last to the first.

Example

Convert the following to the specified bases.
(a) 327 to base six   (b) 7985 to base eight.

Solution

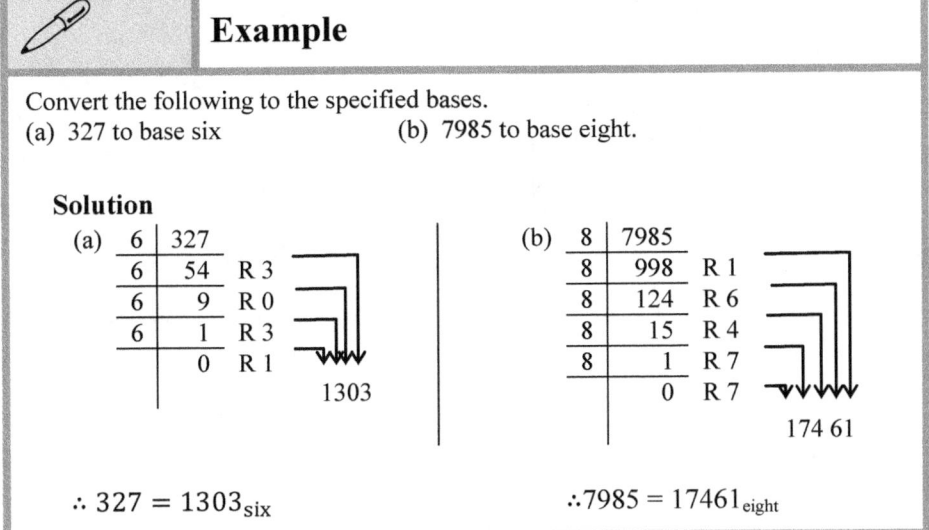

(a)

6	327	
6	54	R 3
6	9	R 0
6	1	R 3
	0	R 1

1303

(b)

8	7985	
8	998	R 1
8	124	R 6
8	15	R 4
8	1	R 7
	0	R 7

174 61

$\therefore 327 = 1303_{six}$ $\therefore 7985 = 17461_{eight}$

Skill Building Exercise 2:8

Convert the following base ten numerals to the bases specified.

(1) 789 to base two   (2) 4899 to base eight   (3) 316 to base six
(4) 3057 to base four   (5) 1349 to base three   (6) 537 to base nine
(7) 29 to base two   (8) 8403 to base six   (9) 792 to base eight
(10) 2740 to base seven

(iii) Converting Non-Denary to Non-Denary

To convert a number from a non-denary base to another non-denary base, first convert the number to base ten and then change the result to the destination base.

 Example

Convert 276_{eight} to base four.

Solution
Changing 276_{eight} to base ten
$$276_{eight} = 2 \times 8^2 + 7 \times 8 + 6 = 128 + 56 + 6 = 190_{ten}$$

Changing 190_{ten} to base 4

(a)

4	190	
4	47	R 2
4	11	R 3
4	2	R 3
	0	R 2

23 32

$$\therefore 276_{eight} = 2332_{four}$$

Skill Building Exercise 2:9

In the following table, the column A is a numeral in a given base. Convert the numeral to a numeral in the base specified in column B.

	Column A	Column B
1.	142_{five}	seven
2.	11011_{two}	eight
3.	45631_{seven}	five
4.	421_{six}	four
5.	7481_{nine}	three
6.	1322_{four}	nine
7.	3472_{eight}	seven
8.	2416_{seven}	six
9.	43021_{six}	four
10.	2102_{three}	seven
11.	3214_{five}	eight
12.	2311_{four}	six
13.	24162_{seven}	four
14.	1011101_{two}	nine
15.	7104_{eight}	three

 ## Integration Activity 2

1. In a certain traditional society, kola nuts are counted in fives as follows:

    5 kola nuts   = 1 hand
    5 hands   = 1 heap
    5 heaps   = 1 basket
    5 baskets   = 1 bag
    5 bags   = 1 load
    5 loads   = 1 ban

    (i) How many kola nuts are there in 3 baskets?

    (ii) How many hands of kola nuts are there in 2 bans?

    (iii) During a long vacation, you decide to be a sales agent of a kola nut dealer who gives kola nuts to sales agents at 4000 francs per bag to sell at 5000 francs per bag. At the end of the period, you sell 4 bans 3 loads and 4 bags.

    (a) How many bags did you sell altogether?

    (b) At how much did you sell the kola nuts?

    (c) How much will you pay to the kola nut dealer?

    (d) What is your profit on the sale of the kola nuts collected?

2. The following shows the partial results of the 2011 Presidential elections in Cameroon. Classify the candidates in order of merit and declare the results of the elections.

Candidate	Number of votes
Adamu Ndam Njoya	83,860
Paul Biya	3,772,527
Garga Haman Adji	155,348
Ni John Fru Ndi	518,175

Source: Wikipedia, the free Encyclopedia.

3. List ten different identity card numbers issued in different regions of your country.
4. List ten different car numbers issued in different regions of your country.
List four different telephone numbers by different service providers in your country.

 Multiple Choice Exercise 2

1. The correct arrangement of the numbers in increasing order of magnitude is:
   [A] 2, 6, 9, 13, 17 [B] 6, 13, 2, 17, 9 [C] 13, 2, 17, 9, 6 [D] 17, 13, 9, 6, 2

2. It is true to say that:
   [A] $38 \leq 15$ [B] $15 \geq 38$ [C] $38 < 15$ [D] $38 > 15$

3. 180×7
   [A] 1900   [B] 1705   [C] 1260   [D] 2000

4. The result of   43
                  $\times 63$ is:
   [A] 1500   [B] 8250   [C] 1845   [D] 2709

5. In $5 + 0 = 5$, the property that applies is:
   [A] The commutative property of zero
   [B] The associative property of zero
   [C] The distributive property of zero
   [D] The additive identity property of zero

6. In $2 \times 0 + 7 = 7$, the property of zero that applies is:
   [A] The multiplicative property   [B] The additive property
   [C] The distributive property   [D] The identity property

7. In $3 \times 1 + 2 = 5$, the property of one that applies is:
   [A] The distributive property   [B] The additive property
   [C] The multiplicative property   [D] The identity property

8. In $(6 \times 2) \times 50 = 6 \times (2 \times 50)$, the property used is:
   [A] The commutative law of multiplication

[B] The associative law of multiplication
[C] The distributive law of multiplication
[D] The multiplicative property of numbers

9. $(73 + 25) + 75 = 73 + (25 + 75)$. The property applied is:
[A] The associative law of addition [B] The commutative law of addition
[C] The distributive law of addition [D] The addition property of numbers

10. In $3 \times 7 = 21$, the name given to 3 is:
[A] the multiplier [B] the multiplicand [C] the minuend [D] the dividend

11. In base five 75_{ten} is the same as:
[A] 300_{five} [B] 400_{five} [C] 500_{five} [D] 600_{five}

12. When expressed as a binary number 27_{ten} is equal to:
[A] 1110_{two} [B] 1111_{two} [C] 11011_{two} [D] 1001_{two}

13. The denary (base ten) number 37, written in binary (base two) is:
[A] 100011_{two} [B] 100111_{two} [C] 100001_{two} [D] 100101_{two}

14. The denary number 39 written in binary is:
[A] 100011_{two} [B] 100101_{two} [C] 100111_{two} [D] 10001_{two}

15. The value of the decimal number 89 as a binary number is:
[A] 101101_{two} [B] 1011001_{two} [C] 1001001_{two} [D] 1001101_{two}

16. 35 in base two is:
[A] 1000_{two} [B] 10011_{two} [C] 100011_{two} [D] 110010_{two}

17. The equivalent of 11111_{two} in base ten is:
[A] 9 [B] 17 [C] 19 [D] 31

18. 101101_{two} in expanded form is the same as:
[A] $1 \times 2^{-6} + 0 \times 2^{-5} + 1 \times 2^{-4} + 1 \times 2^{-3} + 0 \times 2^{-2} + 1 \times 2^{-1}$
[B] $1 \times 2^{-5} + 0 \times 2^{-4} + 1 \times 2^{-3} + 1 \times 2^{-2} + 0 \times 2^{-1} + 1 \times 2^{0}$
[C] $1 \times 2^{5} + 0 \times 2^{4} + 1 \times 2^{3} + 1 \times 2^{2} + 0 \times 2^{1} + 1 \times 2^{0}$
[D] $1 \times 2^{6} + 0 \times 2^{5} + 1 \times 2^{4} + 1 \times 2^{3} + 0 \times 2^{2} + 1 \times 2^{1}$

19. As a number in base ten 321_{five} is equal to:
[A] 85_{ten} [B] 86_{ten} [C] 32_{ten} [D] 43_{ten}

20. 42_{five} is equivalent to:
[A] 21_{ten} [B] 10101_{two} [C] 112_{four} [D] 212_{three}

21. 24 is equivalent to:
[A] 110_{two} [B] 112_{three} [C] 11000_{two} [D] 220_{ten}

22. The possible binary number in the following is:
[A] 112 [B] 101 [C] 102 [D] 211

23. 111_{four}, 22_{eight}, 11011_{two} in ascending order of size is:
[A] 111_{four}, 22_{eight}, 11011_{two} [B] 22_{eight}, 111_{four}, 11011_{two}
[C] 11011_{two}, 22_{eight}, 111_{four} [D] 11011_{two}, 111_{four}, 22_{eight}

31

Topic 3

INTEGERS

Objectives

At the end of this topic, the learner should be able to:

1. Recognize the need to extend to extend the set ℕ of natural numbers to the set ℤ of integers.
2. Read and write negative integers (and not minus integers) correctly.
3. Compare integers using the symbols < and >.
4. Represent integers on the number line.
5. Identify the relative position of integers on the number line.
6. Add and subtract integers.
7. Multiply and divide integers with results in ℤ.
8. Perform with maximum flexibility mixed operations involving the four basic arithmetic operations (+, −, ×, ÷) with results in ℤ.

3.1   Need to Extend the Set of Natural Numbers

> **? Brainstorming Exercise**
>
> 1. Mr. Tenjong saved 40,000 FRS in his bank account and borrowed 30,000
>   FRS from the bank on the 01/03/2011.
>   (a) What amount does he have in his account as of this date?
>   (b) On the 01/04/2011, he borrows 30,000 FRS again from the bank. What
>   amount does he now have in his account?
>   (c) How should the bank write this amount to show that Mr. Tenjong
>   owes?
> 2. The freezing point of water is 0° C. Water at 0° C is put into a refrigerator
>   and its temperature goes below the freezing point.
>   (a) Give two likely values of the new temperature.
>   (b) How can we spare ourselves the stress of using adjectives and long
>   phrases such as 'above the freezing point' and 'below the freezing
>   point'?
> 3. The highest location on the continents is the summit of Mount Everest,
>   which is 8,850 m above sea level, and the lowest elevation on the continents
>   is the Dead Sea between Israel and Jordan, which is 408 m below sea level.
>   Without using the words 'above sea level' and 'below sea level' state the
>   location of the summit of Mount Everest and the Dead Sea between Israel
>   and Jordan distinguishing them as places which are above or below the
>   earth's surface?

Mr. Tenjong's financial statement will be similar to the following.

Amount saved on 01/03/2011   = 40,000 FRS
Amount borrowed on 01/03/2011   = 30,000 FRS
Balance in account on 01/03/2011 = Amount saved −Amount borrowed
                                  = 40,000 FRS − 30,000 FRS
                                  = 10,000 FRS

Balance in account on 01/03/2011 = Amount in account−Amount borrowed
                                  = 10,000 FRS−30,000 FRS
                                  = ? FRS

One may be tempted to say that the amount in his account is 20,000 FRS, but
instead Mr. Tenjong owes the bank 20,000 FRS. If the bank gives him 20,000
FRS, the bank will be at a loss. In this case what the bank normally records is
−20,000 FRS.

Notice the use of words or phrases such as save, owe, above freezing point,
below freezing point, above sea level, below sea level in the above instances to
clarify the situation. The following table shows the way Mathematicians qualify
adjectives such as those used above. They even use them in calculations without
having to bother whether it is above or below and their answers come out just
right.

40,000 FRS saved	+40,000 FRS
30,000 FRS owed	−30,000 FRS
8,850 m above sea level	+8,850 m
408 m below sea level	−408 m
5° C above freezing point	+5° C
5° C below freezing point	−5° C

A number with a + sign in front is called a positive number.

A number with a − sign in front is called a negative number.

Positive and negative whole numbers are called **integers**. Zero is an integer and though can be written as ± 0, the sign is conventionally omitted.

With this new set of numbers, subtraction such as $11 − 13$, $6 − 9$ which at first was impossible in the set of natural numbers can be done.

Care must be taken to read and distinguish between the signs "+" and "−" when used as operations and when used as directed number signs.
+ 5 is read 'positive five' <u>not</u> "plus five'

− 5 is read 'negative five' <u>not</u> "minus five'

−2+ (−3) is read 'negative 2 plus negative 3'

+2− (+3) is read 'positive 2 minus positive 3'

3.2   The Set $\mathbb{Z}$ of Integers

The set of integers is denoted by $\mathbb{Z}$ and is the set of all positive and negative whole numbers including zero.

Thus,

$$\mathbb{Z} = \{0, \pm 1, \ \pm 2, \ \pm 3, ...\}$$

We can partition the set of integers into three sets namely;

The set of positive whole numbers denoted by $\mathbb{Z}^+ = \{+1, +2, +3, ...\}$, the set of negative whole numbers denoted by $\mathbb{Z}^- = \{... , -3, -2, -1\}$ and the set $\{0\}$ containing the element 0.

Real life Examples

The following are some practical applications of Integers.

Banking - Saving and Loans
In banking, savings and deposits are regarded as positive while loans and withdrawals are regarded as negative. Often, thinking of positive number as savings and negative numbers as loans makes operations with integers easier.

Temperatures
Temperatures above freezing point are *positive*. Hence (+) indicates a temperature which is above the freezing point while (−) indicates a temperature which is below the freezing point. (Figure (a) below).

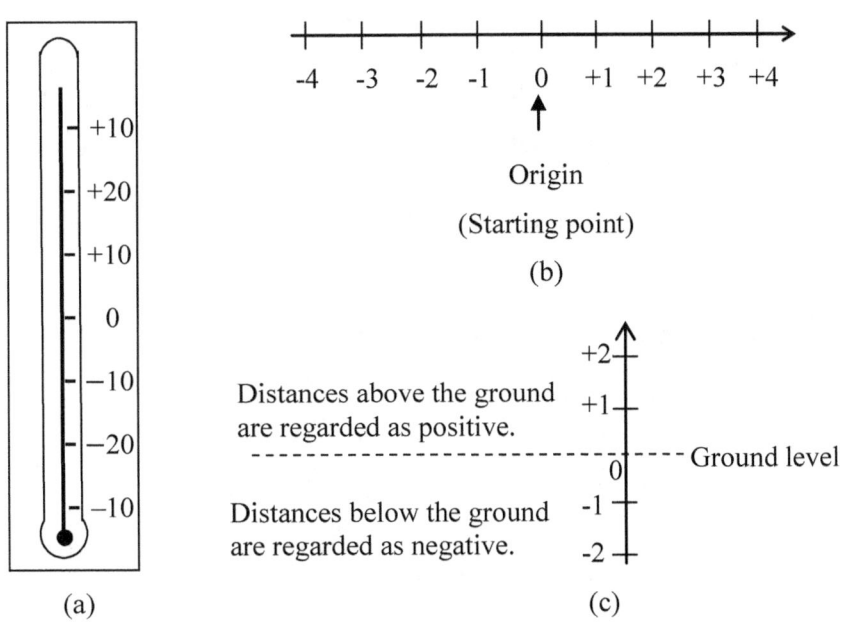

Distances
Distances to the left (or backward) are considered negative (−) while distances to the right (or forward) are considered positive (+). (Figure (b) above).
Distances above the earth's surface are considered negative (+) while distances below the earth's surface are considered positive (−). (Figure (c) above).

3.3   The Integral Number Line

Numbers are often represented on a horizontal or vertical line called the **number line** as shown below. On a horizontal number line the larger number is always to the right. The point zero (0) is called the origin and is the starting point.

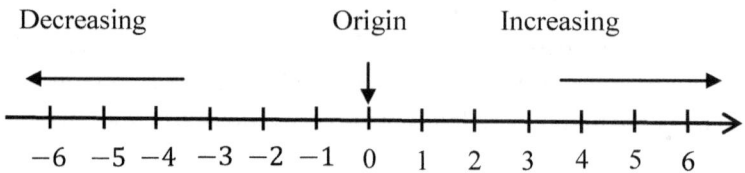

3.4   Comparing Integers

Recall that, < means 'is less than' and > means 'is greater than'.

≤ means 'is less than or equal to' e.g. $-2 \leq 5$ means '−2 is less than or equal to 5'. This is true since −2 is less than 5.

≥ means 'is greater than or equal to' e.g. $9 \geq -5$ means '9 is greater than or equal to −5'. This is true since 9 is greater than −5.

From the figure above, we should appreciate that $-2 > -3, 0 > -2, -3 < +1$ etc.

⚒ Skill Building Exercise 3:1

1. Read aloud the following.
    (a) −7   (b) +13   (c) +10   (d) −31
    (e) +3 + (+7)   (f) −5 − (−2)   (g) −10 + (− 2)
    (h) +8 − (+10)   (i) +14 + (−1)   (j) −9 − (+6)
2. If +70 means 70 francs gained, state the meaning of:
    (a) +30   (b) −50   (c) −10   (d) +20
3. If −3 means 3 points down, state the meaning of:
    (a) −12   (b) +6   (c) −7   (d) +9
4. If −5 means 5 days ago, state the meaning of:
    (a) −6   (b) +2   (c) −4   (d) +3
5. If +6 means 6 m above sea level, state the meaning of:
    (a) −144   (b) +27   (c) +35   (d) −17
6. If −500 means 500 francs spent, state the meaning of:
    (a) −10   (b) +35   (c) −20   (d) +200
7. Which is greater?
    (a) −5 or 0   (b) +7 or −2   (c) −9 or −11   (d) +5 or +8
8. Compare the following using <, > or =

   (a) −2   −1   (b) −10   0   (c) 0   −5
   (d) +7×0   7   (e) −7   −8   (d) +9 + 0   9×1

9. Arrange in order of increasing magnitude the numbers:
   −9, −11, −, +2,−3, +13, −14, +5,+3
10. Arrange the following in order of increasing magnitude. +5−2, +9, +7−5,−4
11. Mark the following points on a horizontal number line. 3, −1, −4, 0
12. On a horizontal number line represent the numbers described in each of the following sets
    (a) The set of natural odd numbers less than 10.
    (b) The set of even integers less than −1 and greater than −10.
13. Suggest as many meanings as possible of the following that occur on a bank statement.   (a) −17,000   (b) 13, 000
14. State the absolute value of each of the following.
    (a) +20   (b) −13   (c) +58   (d) −356
15. Fill the blank spaces with the symbol >, < or =.
    (a) 4 ___12   (b) −7___5   (c) 14 ___ − 6   (d) −8 ___ − 11
16. Answer true or false.
    (a) 2 < 8   (b) −5 > 2   (c) −5 ≤ 5   (d) 7 > −4

3.5 Operations With Integers

Many students at the early stage find it very difficult to manipulate integers. However, this can be facilitated by the use of the number line method, the savings and loans method or the algebra tiles method. After enough skills and facility have been developed, the methods will be unnecessary and can be aborted at that stage.

Algebra Tiles

We can label algebra tiles as ⊕ and ⊖, where ⊕ = +1 and ⊖ = −1. Algebra tiles are useful in creating models for manipulating integers. In addition and subtraction, the sum of ⊕ and ⊖ gives ⊕⊖ whose sum is zero. ⊕⊖ is called a zero pair.

Addition and Subtraction of Integers

Adding Two Positive Numbers

 Example

Evaluate +5 + (+3) using
(a) A number line. (b) The savings and loans analogy. (c) Algebra tiles.

(a) **Using the number line.**
From 0, count 5 units to the right and from 5 count 3 units to the right.

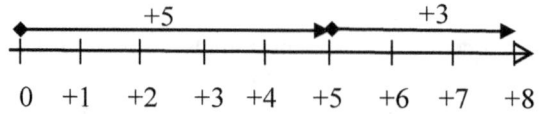

$\therefore +5 + (+3) = +8$

(b) **Using the savings and loans analogy**

If one saves 5 and saves 3, the net amount in his account will be 8.
$\therefore +5 + (+3) = +8$

(c) **Using algebra tiles**

⊕ ⊕ ⊕ ⊕ ⊕ ⊕ + ⊕ ⊕ ⊕ = ⊕ ⊕ ⊕ ⊕ ⊕ ⊕ ⊕ ⊕

$\therefore +5 + (+3) = +8$

Generally, *the sum of two or more positive numbers is a positive number.*

Adding two negative numbers

 Example

Evaluate $-2 + (-3)$ using the three methods introduced in Example 4:2.

(a) **Using the number line.**
From 0, count 2 units to the left and from -2 count 3 units to the left.

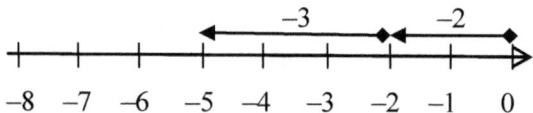

$\therefore -2 + (-3) = -5$

(c) **Using algebra tiles**

⊖ ⊖ + ⊖ ⊖ ⊖ = ⊖ ⊖ ⊖ ⊖ ⊖

$\therefore -2 + (-3) = -5$

Generally, *the sum of two or more negative numbers is a negative number.*

Hence, to add numbers with a common sign,

Write down the common sign of the numbers before the sum of the numbers.

Note that a + sign which stands for positive is usually omitted. Thus, except for emphasis, + 5, is written as 5 and +5 + (+3) simply as 5 + 3.

Adding a Positive number and a Negative number

 Example

1. Find the value of $+5 + (-2)$ using the three methods.
    (a) **Using the number line.**
        From +5 count 2 units to the left.

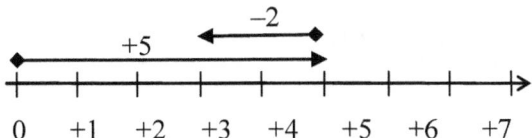

    $\therefore +5+(-2) = +3$

    (b) **Using the savings and loans analogy**
        If one saves 5 and loans 2, the net amount in his account will be 3.
        $\therefore +5 + (-2) = +3$

    (c) **Using algebra tiles**

        $\oplus \oplus \oplus \oplus \oplus + \ominus \ominus = \oplus \ominus \; \oplus \ominus \quad + \; \oplus \oplus \oplus$
        (Zero pairs = 0)

    $\therefore +5 + (-2) = +3$

2. Find $-4 + (+3)$ using the three methods.

    (a) **Using the number line.**
        From -4, count 3 units to the right.

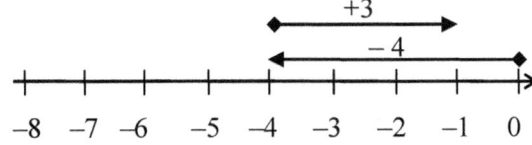

    $\therefore -4 + (+3) = -1$

    (b) **Using the savings and loans analogy**
        If one loans 4 and saves 3, he owes 1.

$$\therefore -4 + (+3) = -1$$

(c) **Using algebra tiles**

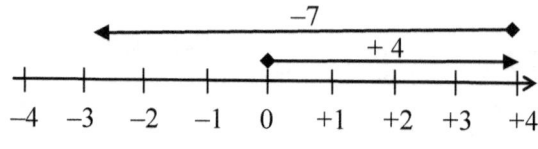

(Zero pairs = 0)

$$\therefore -4 + (+3) = -1$$

3. Evaluate +4 + (–7) using the three methods.
   (a) **Using the number line.**

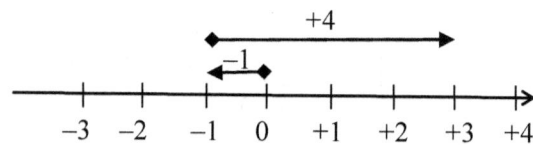

$$+4 + (-7) = -3$$

(b) **Using the savings and loans analogy**
A saving of 4 and a loan of 7 is a debt of 3.
$$\therefore +4 + (-7) = -3$$

(c) **Using algebra tiles**

$$⊕⊕⊕⊕ + ⊖⊖⊖⊖⊖⊖⊖ = \underline{⊕⊖} \; \underline{⊕⊖} \; \underline{⊕⊖} \; \underline{⊕⊖} + ⊖⊖⊖$$

(Zero pairs = 0)

$$\therefore +4 + (-7) = -3$$

4. Determine the value of –1 + (+4).
   (a) **Using the number line.**

$$\therefore -1 + (+4) = +3$$

(b) **Using the savings and loans analogy**
A loan of 1 and a saving of 4 is a saving of 3.
$$\therefore -1 + (+4) = +3$$

(c) **Solution by the algebra tiles method**

$$⊖ + ⊕⊕⊕⊕ = \underline{⊖⊕} + ⊕⊕⊕$$

$$\therefore -1 + (+4) = +3$$

Notice that in adding a positive and negative number the sign of the result is always the sign of the number with the larger absolute value.

Generally, *to add two numbers with unlike signs, subtract the number with the smaller absolute value from the one with the larger absolute value and retain the sign of the number with the larger absolute value.*

To add more than two integers, first add the numbers with like sign.

 Example

1. Evaluate $+5 + (-6) + (-8)$

   Solution
   $+5 + (-6) + (-8) = +5 + (-14) = -(|-14|-|5|) = -9$

2. Compute $-7 + (-6) + (-4)$

   Solution
   $-7 + (-6) + (-4) = -13 + (-4) = -17$

   When a $(+)$ sign and a $(-)$ sign occur together, the $(-)$ sign takes preference.

3. What is the value of $3 + (-4) - (+5)$?

   Solution
   $3 + (-4) - (+5) = 3 - 4 - 5 = 3 - 9 = -(9-3) = -6$

4. Evaluate $-47 + 58 + (-5)$.

   Solution
   $-47 + 58 + (-5) = -47 + (-5) + 58 = -52 + 58 = 58 - 52 = -6$

 Skill Building Exercise 3:2

Evaluate the following.
1. $(+20) + (-7)$
2. $(-17) + (+3)$
3. $(-5) + (-15)$
4. $(+5) + (-15)$
5. $(+8) + (+20)$
6. $(-8) + (+8)$
7. $(+5) + (+3) + (+12)$
8. $(-1) + (-3) + (-2)$
9. $(+23) + (-13) + (+12)$
10. $(+5) + (-8) + (+17)$
11. $(+12) + (-12) + (+19)$
12. $(-42) + (+42) + (-85)$
13. $(+11) + (+3) + (-9) + (-8)$
14. $(+23) + (-16) + (-22) + (+16)$

The Additive Inverse of a number

The counterpart of a number such that their sum is zero is called its **additive inverse**. For instance, since the sum of -4 and $+4$ is 0, -4 is said to be the additive inverse of $+4$ and vice versa.

Example

State the additive inverse of each of the following.
(a) +17   (b) − 15   (c) 30   (d) − 48

Solution
(a) −17   (b) 15   (c) −30   (d) 48

Subtraction of Integers

Subtracting an integer is the same as adding its additive inverse. Equally we saw that $-(+) = -$ and $+(-) = -$. We can apply these rules to facilitate subtraction of integers.

 Example

Evaluate (a) 13 − (+5)   (b) +15− (− 18)

Solution
(a) $13 - (+5) = 13 - 5 = 8$ (b) $+15 - (-18) = 15 + (+18) = 15 + 18 = 33$

 Skill Building Exercise 3:3

1. State the additive inverse of each of the following.
     (a) +25   (b) − 53   (c) 42   (d) − 74
2. Evaluate the following
     (a) (+20)−(−7)   (b) (−17)−(+3)   (c) (−5)−(−15)
     (d) (+5)− (−15)   (e) (+8)−(+20)   (f) (−8)−(+8)
     (g) (+5) − (+3) − (+12)   (h) (−1) − (−3) − (−2)
     (i) (+23) − (−13) − (+12)   (j) (+5) − (−8) − (+17)
     (k) (+12) − (−12) − (+19)   (l) (−42) − (+42) − (−85)
     (m) (+11) − (+3) − (−9) − (−8)   (n) (+23) − (−16) − (−22) − (+16)

Multiplication of Integers

Multiplication is repeated addition. Thus;

$$(+4) \times 5 = (+4) + (+4) + (+4) + (+4) + (+4) = +20 = 20$$

$$(-4) \times 5 = (-4) + (-4) + (-4) + (-4) + (-4) = -20$$

We can use algebra tiles to model the product − 4 × 5 as follows.

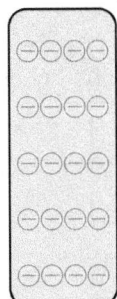

There are 5 groups of − 4. By counting all these 5 groups of − 4, it can clearly be seen that − 4×5 = − 20.
Since multiplication is commutative
$$-4 \times 5 = 5 \times (-4) = -20$$

Generally, *the product of a negative number and a positive number is negative.*

 Example

Evaluate the following.   (a) − 7× 3   (b) 6 ×(−8)

Solution
Evaluate the following.   (a) − 7× 3 = −21   (b) 6 ×(−8) = − 48

? **Brainstorming Exercise**

What is the sign of the product of two negative numbers?

Consider − 4 × (− 5).

Since − 4 = − (+4), − 4 × (− 5) = − (+4) × (− 5)) = − (− 20)

− (− 20) is the opposite of count 20 steps to the left i.e. count 20 steps to the right. Thus, − (− 20) = +20.

Therefore, − 4× (−5) = +20

Generally, *the product of two negative real numbers is a positive real number.*

 Example

Evaluate (a) −13(−5)   (b) (−15)(− 8)

Solution
(a) −13(−5) = 65   (b) (−15)(− 8) =120

Summary on Multiplication

The table below is the summary of the sign of the product of two real numbers.

Like signs	Unlike signs
$(+)(+) = (+)$ and $(-)(-) = (+)$	$(+)(-) = (-)$ and $(-)(+) = (-)$

The general rule for multiplying two or more integers is that if the number of negative signs is even the result is positive and if the number of negative signs is odd, the result is negative.

Odd Number of negative signs	result is Negative $(-)$
Even Number of negative signs	result is Positive $(+)$

✖ Skill Building Exercise 3:4

1. Multiply the following.
   (a) $(+5)(-3)$
   (b) $(-4)(+3)$
   (c) $(-5)(-3)$
   (d) $(+5)(-6)$
   (e) $(+8)(+2)$
   (f) $(-8)(+8)$

2. Multiply the following.
   (a) $(+5)(+3)(+1)$
   (b) $(-1)(-3)(-2)$
   (c) $(+3)(-1)(+2)$
   (d) $(+5)(-8)(+1)$
   (e) $(+2)(-1)(+4)$
   (f) $(-2)(+4)(-5)$
   (g) $(+1)(+3)(-4)(-2)$
   (h) $(+3)(-1)(-2)(+4)(-3)$

3.6   Mixed Operations Involving Integers

In problems which involve integers and contain a mixture of the operations $+$, $-$ and $\times$, $\times$ should be done first before $+$ and $-$. In other words, the order of operation is multiplication, addition and subtraction. The mnemonic **MAS** may help to bring this fact to mind.

Example

1. Find the value of $(+2) \times (-4) + (-7) - (+8) \times (-2)$

Solution
$$(+2) \times (-4) + (-7) - (+8) \times (-2) = (-8) + (-7) + 16$$
$$= -15 + 16 = +1$$

2. Evaluate $(-8) \times (-2) - (+4) + (-3) - (+2) \times (-3)$

Solution
$$(-8) \times (-2) - (+4) + (-3) - (+2) \times (-3) = 16 - (+4) + (-3) + 6$$
$$= 16 - (+4) + (-3) + 6 = -7$$

Skill Building Exercise 3:5

Evaluate the following

1. $(+7) + (-3) \times (+6) - (+6)$
2. $(-5) + (+8) \times (-3)$
3. $(+9) + (+2) \times (-3) - (-6)$
4. $(-3) + (+6) \times (-12)$
5. $(+3) \times (-5) + (-9) - (+6) \times (-3)$
6. $(-2) \times (-3) - (+2) + (-8) - (+3) \times (-6)$

Integration Activity 3

A businessman made a gain of 2300 francs on Monday, a gain of 3600 francs on Tuesday, a loss of 1200 francs on Wednesday, a gain of 4100 francs on Thursday, a loss of 2700 francs on Friday, and a loss of 1800 francs on Saturday.
(a) Determine his net profit or loss that week.
(b) Given that his balance brought forward from the previous week was 876275 francs, what will be his net income at the end of this week?

Multiple Choice Exercise 3

1. The greatest among the following is:
    [A] −2   [B] −4   [C] −6   [D] −8
2. When evaluated −7+5−2+11−8 equals:
    [A] +1   [B] −1   [C] +33   [D] −33
3. The value of −6−(−6) is:
    [A] −12   [B] 0   [C] 12   [D] 36
4. (−4)(−3) is equal to:
    [A] −7   [B] 7   [C] −12   [D] 12
5. (−2)×(−3) is equal to:
    [A] −6   [B] 6   [C] −5   [D] 5
6. The value of (+4)(−2) − (−2) is:
    [A] −6   [B] −4   [C] +4   [D] +6
7. −120 + 83 equals:
    [A] −47   [B] −43   [C] −37   [D] 37
8. The result of (−1)(−2)(+3)(+4) is:
    [A] +18   [B] −24   [C] −18   [D] +24
9. What is the value of (−6)×(−7)×(−3) is:
    [A] −39   [B] 39   [C] −5   [D] −126
10. 27÷(−3) is equal to:
    [A] 9   [B] −9   [C] −7   [D] 7
11. As an integer, 32°F below zero degrees is:
    [A] −31°   [B] 32°   [C] −32°   [D] 31°
12. The additive inverse of the integer −3 is:
    [A] 1   [B] 3   [C] −3   [D] 0
13. The absolute value of 5 is:
    [A] 5   [B] −5   [C] −1   [D] 0
14. The correct order of the integers 8, 15, −1, 6, −6 from least to greatest is:
    [A] −6, −1,6,8,15   [B] −1, −6,15,8,6
    [C] 8,6, −6, −1,15   [D] 15,8,6, −1, −6
15. The sum 1 and |− 5| is:
    [A] 6   [B] 4   [C] −4   [D] −6
16. The difference −5 − (−2) is:
    [A] 3   [B] −3   [C] −7   [D] 7
17. A fish was swimming at 180 m below sea level. Then it descended to 315 m below sea level. As an integer in meters, the change in the fish's depth is:
    [A] 135   [B] −135   [C] −495   [D] 495
18. 8 × (−4) is:
    [A] −16   [B] −32   [C] 32   [D] 16
19. Using the following table, Tangwe's profit or loss for the month of January is:

Income and Expenses for Tangwe

Month	Income	Expenses
January	1,817 FCFA	−2,338 FCFA
February	2,271 FCFA	−2,315 FCFA
March	3,243 FCFA	− 1,530 FCFA
April	3,929 FCFA	−1,167 FCFA
May	3,477 FCFA	−1,101 FCFA
June	3,077 FCFA	−834 FCFA

[A] −4155 FCFA [B] −4155 FCFA [C] −521 FCFA [D] 521 FCFA

20. The result after evaluating $5 \times 4 - 8 \times 6$ is:
   [A] 28   [B] −28   [C] −72   [D] 72

21. $4 \times 6 - 10 \times 5 + 2$ after simplification gives:
   [A] 72   [B] −72   [C] 24   [D] −24

22. As air moved into a region where the temperature was originally 25° C, the temperature began falling at a steady rate of 4°C per hour. The temperature after 9 hours would be:
   [A] −7°C   [B] −11°C   [C] 36°C   [D] −15°C

23. Anye owns a small business. There was a loss of 140 FCFA on Thursday and a loss of 120 FCFA on Friday. On Saturday there was a loss of 110 FCFA, and on Sunday there was a profit of 160 FCFA. The total profit or loss for the four days is:
   [A] 310 FCFA loss   [B] 250 FCFA profit
   [C] 210 FCFA loss   [D] 530 FCFA profit

Topic 4

TIME AND TEMPERATURE

Objectives

At the end of this topic, the learner should be able to:

1. Read, write and narrate historical events using BC (Before Christ) and AD (Anno domini).
2. Describe time of the day using a.m. and p.m.
3. Convert time from a 12 hour system to a 24 hour system and vice versa.
4. State and use time in one time zone, given time in another time zone.
5. Convert from one unit of time to another.
6. Determine when to use smaller units such as years or less or when to use larger units of time such as decade, century, millennium etc.
7. Read and record temperatures in the Celsius and Fahrenheit scale.
8. Convert temperature written in Celsius to Fahrenheit and vice versa.

TIME

4.1 Historical Time

Historical time is often written with the suffixes B.C. which means Before Christ (was born) or A.D., which means Anno Domini in Latin, translated in English as in the year of our Lord. The number line in figure illustrates that 100 B.C. when Julius Caesar was born is earlier than the year 0 estimated as the birth year of Jesus Christ.

Cameroon Had its independence 1961 years after Jesus Christ was born.

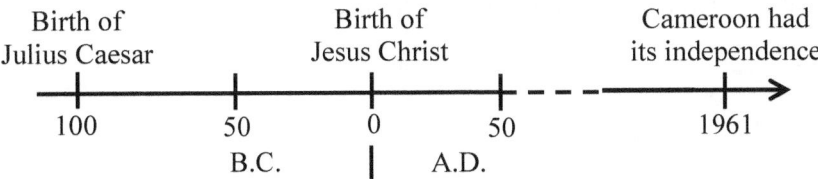

Birth of Julius Caesar	Birth of Jesus Christ	Cameroon had its independence
100	50 0 50	1961
	B.C. A.D.	

Julius Caesar

In 45 BC Julius Caesar established the first Calendar based on a solar year of 365 and a quarter days. In this calendar, the accumulated fourth of a day were dealt with every four years by a convention called the leap year, which is still observed today. In addition the so-called Julian calendar determined the order of the months and the days of the week, which form the bases of the modern calendar. This Calendar has stood the test of time as it is still the universal one used today.

 Competency Based Exercise 4:1

The following shows some major events in Egyptian History. Arrange these events in the order in which they occurred.
(a) In 332 BC, Macedonian king Alexander the Great conquered Egypt.
(b) In 1798 AD Napoleon Bonaparte of France invaded Egypt.
(c) In 3500 BC the Egyptians created a form of writing called hieroglyphs.
(d) In 641 AD Muslim Arab invaders conquered Egypt.

4.2   Time Units

Below is the relationship between time units.

60 seconds make 1 minute (min)
60 minutes make 1 hour (h)
24 hours make day (d)
7 days make 1 week (wk.)
4 weeks make 1 month (Mth)
12 months make 1 year

10 years make 1 decade
10 decades make 1 century
10 centuries make 1 millennium

Conversion of Time Units

1.   To convert from a larger unit to a smaller one, multiply.
2.   To convert from a smaller unit to a larger one, divide.

 Example

Convert
(1) 40 minutes to seconds   (2) 3 hours to seconds.   (3) 2 days to minutes.
(4) 4 years to hours.   (5) 52560 hours to years.

Solution
(1) 40 minutes = 40 × 60 seconds = 2400 seconds
(2) 3 hours = 3 × 60 minutes = 3 × 60 × 60 seconds = 10800 seconds
(3) 2 days to minutes = 2 × 24 hours = 2 × 24 × 60 minutes = 2880 minutes
(4) 4 years to hours = 4 × 365 days = 4 × 365 × 24 hours = 35040 hours
(5) 52560 hours to years $= \frac{189216000}{60 \times 60 \times 24}$ days $= \frac{189216000}{60 \times 60 \times 24 \times 365}$ years = 6 years

4.3   Time as a Non-Metric S.I. Unit

Time is the only S.I. unit, which does not use the metric system. This means that, the units of time do not follow the decimal system.

 Discussion Exercise

1.   Suppose that, 1 year equals 10 months, how is it going to be like?
2.   Can it be possible?
3.   Give reasons why it can be possible or why it cannot be possible.

Exercise 4:1

1. How many days are there in 20 weeks 4 days?
2. How many seconds are there in 12 minutes 10 seconds?
3. Ayuk took 35 minutes to walk to school. If he reaches school at 7.25 a.m., at what time did he leave his house?
4. A man did a piece of job for 1 year 3 months 3 weeks 2 days. For how long in days did he do the piece of job?
5. Convert 7 days to minutes.

4.4   Clocks and Watches

There are two types of clocks and watches - the 12 hours and the 24 hours clocks and watches.

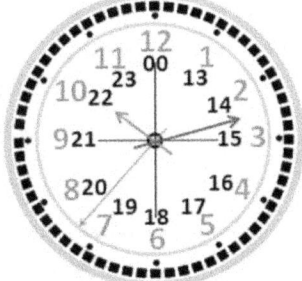

A 24 hour analog clock

A digital clock

A day is 24 hours. Therefore, after 12 noon, 12 hour clocks and watches begin again from 1 instead of going to 13 as the 24 hour clocks. Therefore there is need to distinguish between time before 12 noon and time after 12 noon when we use a 12 hour system. We use the suffix a.m. (ante meridiem) before 12 noon and p.m. (post meridiem) after 12 noon when we read and record time using the 12 hour clock. For instance, if a 12 hour system watch shows 6 O'clock in the evening, the time will be read or written as 6 p.m. but if it shows 6 O'clock in the morning, the time will be read or written as 6 a.m.

The day begins at 12 midnight and ends at 12 midnight. Thus, in the 24 hour system, 12 midnight is considered 0 O'clock or 24 O'clock while 12 noon is just 12 O'clock.

For example if a 24 hour clock or watch reads 19:00, this time will be interpreted as 7 p.m. (19 − 12) in the 12 hour system.

The following table shows more examples how a 12 hour system time can be translated to a 24 hour system time.

12 hour system Time	24 hour system Time
1:40 p.m.	13.40
1:30 a.m.	01:30
8:20 p.m.	20:20
11:15 p.m.	23:15
6:57 a.m.	06:57

 Example

1. Change the following to the 24 hour system time
    (a) 10 p.m.   (b) 2.45 p.m.
2. Change the following to the 12 hour system time.
    (a) 12:00   (b) 16:22

Solution
1. (a) 10 p. m. $= 12 + 10 = 22:00$
    (b) 2.45 p. m. $= 12 + 2.45 = 14:45$
2. (a) $12:00 = 12$ p.m.   (b) $24:00 = 00:00 = 12$ a.m.
    (c) $16:22 = 16:22 - 12:00 = 422$ p. m.

4.5   Time Zones in the World

Due to the rotation of the earth, when it is day in some places, it is night in others. Therefore, for legal, commercial, and social reasons, the earth is divided by longitudes into 24 geographic areas called time zones. Clocks within a given time zone are set to the same time. Each time zone is defined by its distance east or west of Greenwich, England. Time in each of the 12 zones east of Greenwich increases one hour for each zone. Time in each of the 12 zones to the west of Greenwich decreases one hour for each zone. The International Date Line divides the eastern and western time zones. The time difference between each side of the International Date Line is 24 hours. Thus, a traveler heading west across the date line appears to lose one day while a traveler heading east appears to gain a day.

Example

A plane left London at 1:20 p.m. on Tuesday (London time) and arrived in Australia at 5:40 p.m. Wednesday (Australian time). How long did the plane take if Sydney time is 11 hours ahead of London time?

Solution
In terms of London time, time of arrival = Arrival time − 11 hours
$$= 5: 40 - 11 \text{ hours} = 6:40 \text{ a.m.}$$
Therefore time taken = 6:40 a.m. − 1:20 p.m. = 17 hours 20 minutes.

Exercise 4:2

1. State the following as analog time
   (a) 17:25　　　　(b) 23:36　　(c) 14:20　　　　(d) 11:10
2. State the following as digital time.
   (a) 8: 34 p.m.　　(b) 5:56 a.m.　(c) 12:45 p.m.　(d) 12:45 a.m.
3. Calculate the time between
   (a) 10:42 a.m. and 8:12 p.m.　　(b) 10:42 p.m. and 8:12 a.m.
   (c) 8:16 a.m. and 7:20 p.m.　　(d) 22:33 and 9:15
4. What is the meaning of　(a) a.m.?　　　　　(b) p.m.?
5. A football match started at 4 p.m. Spanish local time. A city's time is 13 hours behind Spanish time. At what time did the match start in the local time of the city?
6. A plane left airport A at 8:30 p.m. on Tuesday and landed at airport B at 9:00 p.m. (local time) on the same day. If the time at airport B is 19 hours behind that at airport A. How long did the trip take?
7. When the time is 6:46 p.m. in Dakar, it is 10:46 p.m. in Nairobi. A plane takes 14 hours to travel from Dakar to Nairobi. If the plane leaves Dakar at 2:40 p.m. on Tuesday:
   (a) At what time (local time) will it arrive in Nairobi?
   (b) On what day will it arrive in Nairobi?
8. The following shows a jumbled program of the order in which a minister has to receive some personalities.
   - 1 p.m. The Lord mayor
   - 15:00 the DO.
   - 8 a.m. The commissioner
   - 10.00 The Government delegate
   - 12 noon Political leaders.
   Arrange the program in the order in which the reception will be done.

TEMPERATURE

Temperature is the degree of hotness or coldness of an object. It is measured using a thermometer. The units of temperature are the degree Fahrenheit (°F) and the degree Celsius or Centigrade (°C).

Some significant temperatures worth noting are;

Boiling point of water = 100°C
Normal body temperature = 37°C
Normal room temperature = 20°C
Freezing point of water = 0°C

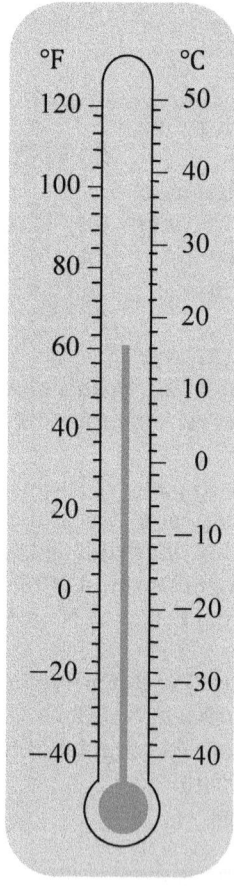

A Fahrenheit and Celsius Thermometer

4.6 Conversion of Temperatures

The formula to convert from degrees Celsius to degrees Fahrenheit is as follows.

$$F = \frac{9}{5}C + 32 \Leftrightarrow C = \frac{(F - 32)5}{9}$$

 Example

Convert (a) 85°C to °F   (b) 149 °F to °C

Solution

(a) $F = \frac{9}{5}C + 32 = \frac{9}{5}(85) + 32 = 185°F$

(b) $C = \frac{(F-32)5}{9} = \frac{(149-32)5}{9} = 65°C$

 Competency Based Exercise 4:2

The thermal requirement for a certain species of chicks is 24°C to 35°C.The fan needs to be put on if the temperature is higher than 35°C and the heater needs to be put on if the temperature is lower than 24°C. You don't have a Celsius thermometer but have a Farenheight thermometer which reads99°F. Which of the heater or the fan will you put on?

 Exercise 4:3

1. Convert to degrees Fahrenheit.
   (a) 60°C   (b) 35°C   (c) 20°C   (d) 40°C   (e) 75°C
2. Convert to degrees Celsius.
   (a) 122°F   (b) 93.2°F   (c) 62.6°F   (d) 199.4°F   (e) 113°F
3. State which of the following is cold, warm or hot.
   (a) 94°C   (b) 35°C   (c) 5°C   (d) 40°C   (e) -8°C   (f) 5°C

Multiple Choice Exercise 4

1. The difference between 3 minutes and 40 seconds is:
   [A] 140 seconds [B] 140 minutes [C] 37 seconds [D] 37 minutes

2. Bih took 35 minutes to go to school. If she arrived at the school at 7:25 a.m., The time she left her house was:
   [A] 6:50 a.m.   [B] 6:55 a.m.   [C] 6:45 a.m.   [D] 6:50 p.m.

3. The number of seconds are there in 12 minutes 10 seconds is:
   [A] 720 seconds [B] 730 seconds [C] 490 seconds [D] 630 seconds

4. The number of days in 30 weeks 5 days is:
   [A] 305 days   [B] 35 days   [C] 180 days   [D] 215 days

5. City A's time is 8 hours ahead of that of city B. If the time at city B is 6:30 p.m. on Tuesday, the time at city A is:
   [A] 2:30 p.m. on Wednesday   [B] 2:30 a.m. on Wednesday
   [C] 2:30 a.m. on Tuesday   [D] 2:30 p.m. on Wednesday

6. The temperature shown by the thermometer below is:
   [A]   6°F   [B]   4°F   [C]   1°F   [D]   0°F

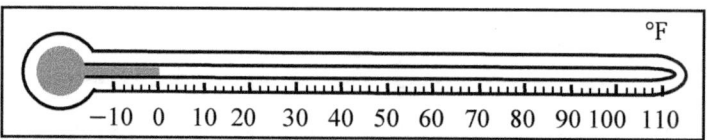

7. The temperature shown by the thermometer below is:

   [A]   74°F   [B]   76°F   [C]   84°F   [D]   86°F

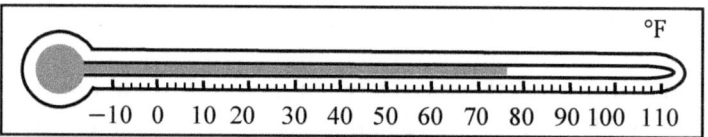

8. The rise in temperature is:

    Previous Temperature

    Present Temperature

    [A]   −50°F   [B]   50°F   [C]   −30°F   [D]   30°F

9. The fall in temperature is:
    [A]   −40°F   [B]   40°F   [C]   −70°F   [D]   70°F

    Previous Temperature

    Present Temperature

Topic 5

NUMBER PATTERNS

Objectives

At the end of this topic, the learner should be able to:

1. Make triangular, square or rectangular dot patterns of numbers in $\mathbb{N}^*$.
2. List the sets of triangular, square and rectangular numbers.
3. Find the factors of a number in $\mathbb{N}$.
4. Identify odd and even numbers among the first ten triangular, square and rectangular numbers.
5. Identify and recognize prime and composite numbers.
6. Express numbers as a product of prime factors.
7. Relate that even numbers can be represented by rectangular patterns of 2 by $n \in \mathbb{N}^*$ but odd numbers cannot.
8. List the factors and multiples of a given natural number.
9. Find the common factors and common multiples of a given natural numbers.
10. Find the HCF and LCM of at most four natural numbers.
11. Express simple perfect squares/cubes in index form.
12. Use prime factorization to find the square root and/or cube root of a given natural number.
13. Recognize by inspection and/or verification whether a number is divisible by 2,3,4,5,6,10,12,25,50,100.

5.1 Dot Representation of Numbers

A number of dots or pebbles carefully arranged forms a square, rectangular or triangular pattern.

? **Brainstorming Exercise**

Look at the following different arrangements of 36 dots.

A   B   C   D   E

1.   Which of the patterns is/are rectangular?
2.   Which of the patterns is/are triangular?
3.   Which of the patterns is/are in a square form?
4.   Draw dot patterns to represent the numbers 3 and 4.
5.   What shapes do the numbers 3 and 4 form?
6.   Can 3 dots be arranged to form a rectangle or a square?
7.   Can 4 dots be arranged to form a triangle or a rectangle?
8.   Try to arrange five dots as a triangle, a square or a rectangle. Is it possible?
9.   Make dot patterns for whole numbers from 6 to 15 and hence classify these numbers as triangular, rectangular or square numbers.
10. Can 0 or 1 form squares, rectangles or triangles?

Clearly, 3 dots can only be arranged to form a triangle and 4 dots can only be arranged to form a square. Five dots cannot be arranged to a triangle, a square or a rectangle.

*Natural numbers whose dot patterns are triangles are called **triangular numbers**; those whose dot patterns are squares are called **square numbers** and those whose dot patterns are rectangles are called **rectangular numbers**.*

Though it may not be easy to use dots or pebbles to appreciate that 0 and 1 belong to all the three sets through logical intuition, this can be ascertained as follows;

Since a square has equal sides, 0 and 1 can be considered squares of side 0 and 1 respectively. In addition, since a square is a special rectangle, all squares are rectangles. Therefore, both 0 and 1 are rectangular numbers. It follows that 1 and 0 are equilateral triangles of side 1 and 0 respectively.

 Exercise 5:1

1. (i) By drawing dot patterns classify the numbers from 1 to 60 as triangular numbers, square numbers or rectangular numbers.
   (Note that some numbers belong to more than one group)

   (ii) (a) List the first three rectangular numbers which are square numbers.
       (b) List the first three rectangular numbers which are triangular numbers.
       (c) List the first three triangular numbers which are square numbers.

2. The dots patterns below represent the first three triangular numbers.

   Make dot patterns to represent the next two triangular numbers and state their values.

3. Write out the next row of each of the following patterns and state a rule for the pattern.

   (a)
   $$1 = 1$$
   $$1 + 3 = 4$$
   $$1 + 3 + 5 = 9$$
   $$1 + 3 + 5 + 7 = 16$$

   (b)
   $$0 = 0$$
   $$0 + 1 = 1$$
   $$0 + 1 + 2 = 3$$
   $$0 + 1 + 2 + 3 = 6$$
   $$0 + 1 + 2 + 3 + 4 = 10$$

   (c)
   $$1 - 1 = 0$$
   $$3 - 2 = 1$$
   $$6 - 3 = 3$$
   $$10 - 4 = 6$$

5.2   Even and Odd Numbers

Even numbers are numbers that can be exactly divided by two with no remainder. The first six even numbers are 2, 4, 6, 8, 10, and 12.

Odd numbers are numbers which when divided by 2, leave a remainder of 1. In other words, odd numbers are not divisible by 2. The first six odd numbers are 1, 3, 5, 7, 9, and 11.

Even numbers can be represented as rectangular patterns with two rows or two columns. Odd numbers on the other hand cannot be represented as rectangular patterns with two rows or two columns.

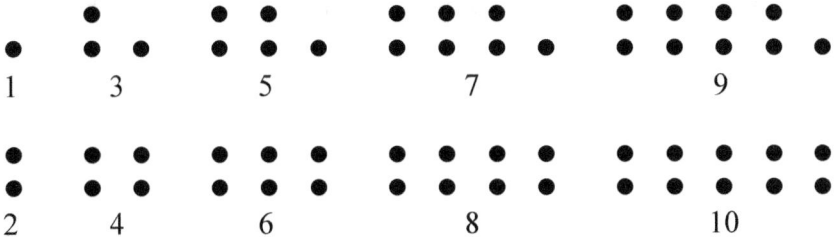

Sum and Difference of Odd and Even Numbers

| | **Investigative Activity** |

In the third column in the table below, write O if the answer is odd and E if it is even.
1. Add pairs of odd numbers.
2. Add pairs of even numbers.
3. Find the difference between odd numbers.
4. Find the difference between even numbers.
5. Find the sum of an odd number and even number.
6. Find the difference between an odd number and even number.

1.	The sum of two odd numbers.
2.	The sum of two even numbers.
3.	The difference between two odd numbers.
4.	The difference between two even numbers.
5.	The sum of an odd numbers and an even number.
6.	The difference between an odd numbers and an even number.

From the above investigations, we can draw the following conclusions.

$$O \pm O = E, \qquad E \pm E = E, \qquad O \pm E = O$$

 Exercise 5:2

1. List the first five even:
   (a) square numbers   (b) rectangular numbers   (c) triangular numbers
2. Let T = All triangular numbers
   S = All square numbers
   R = All rectangular numbers
   The following pattern shows the difference between consecutive members of S.
   The answer of the first has been given.

$$
\begin{aligned}
1 &- 0 &&= 1 \\
4 &- 1 &&= \\
9 &- 4 &&= \\
16 &- 9 &&= \\
25 &- 16 &&= \\
36 &- 25 &&=
\end{aligned}
$$

(a) Copy and complete the pattern.
(b) Name the set of numbers obtained as your results.
(c) As in (a) and (b) make similar patterns for:
   (i) The sum of two consecutive members of T.
   (ii) The difference between two consecutive members of T.
(d) Name the set of numbers obtained in (c)(i) and (ii).

5.3 Factors and Multiples

If the product of two non−zero numbers gives a third number, the third number is said to be a **multiple** of the two numbers, while the two numbers are said to be the **factors** of the third number. Thus,

$$
\underset{\text{factor}}{3} \ \times \ \underset{\text{factor}}{12} \ = \ \underset{\text{multiple}}{36}
$$

36 is a multiple of 3 and 12 while 3 and 12 are factors of 36.

By the definition above it should be clear that every number is a factor of itself and 1 is a factor of every number, since for instance $1 \times 36 = 36$.

Note that apart from the number itself, all the other factors of a number are smaller. Also, apart from the number itself, all the other multiples of a number are larger.

 Example

Write down the set of all the factors of 36.

Solution
Factor of 36 = {1, 2, 3, 4, 6, 9, 12, 18, 36}

Every number is **divisible** by each of its factors. Thus, 36 is divisible by 1, 2, 3, 4, 9, 12, 18 and 36.

 Exercise 5:3

1. List the set of factors of:
   (a) 24   (b) 60   (c) 120   (d) 72   (e) 105   (f) 75
2. List the first 5 multiples of each of the following numbers.
   (a) 2   (b) 4   (c) 5   (d) 7   (e) 8   (f) 12
3. Which of the following are factors of 48? 1, 2, 3, 4, 5, 6, 7, 8, 9
4. Which of the following are multiples of 6?
   18, 20, 21, 22, 24, 26, 27, 28, 30, 32, 34, 35, 36, 38.

5.4   Prime and Composite Numbers

The table below shows some counting numbers and their factors. Notice that some natural numbers have only two factors while others have more than two factors.

Number	Set of factors	Number of factors
1	1	1
2	1,2	2
3	1,3	2
4	1,2,4	3
5	1,5	2
6	1,2,3,6	4
7	1,7	2
8	1,2,4,8	4
9	1,3,9	3
10	1,2,5,10	4

A natural number which has exactly two factors, the number itself and one is

63

called a **prime number**. A **composite number** on the other hand is a number that has more than two factors.

Notice that the number 1 is neither a prime number nor a composite number because it has only one factor, 1 itself.

Notice also that, apart from 2, no other prime number can be arranged as a rectangular pattern. This means that 2 is the only even prime number. Also prime numbers cannot be arranged as triangular, rectangular or square numbers.

 Exercise 5:4

1. Which of the following are prime numbers? 5,27,29,39,24,49,51.
2. Which of the following are composite numbers? 11,17,49,35,24,19.
3. How many prime numbers are there between 1 and 50? List all these prime numbers.
4. List the set of all the factors of 72. How many composite factors, has the number 72?

5.5 Prime Factorization, HCF and LCM

It is possible to express every composite number as a product of a unique set of prime factors. This process of expressing a composite number as a product of prime factors is known as **prime factorization**.

 Example

Express each of the following as a product of prime factors. (a) 30   (b) 24.

Solution
(a) $30 = 2 \times 3 \times 5$ (b) $24 = 2 \times 2 \times 2 \times 3 = 2^3 \times 3$

It is not often as easy as in the above example to express a number as a product of prime factors. Cases that are more difficult require some technique such as the peeling method and the factor tree method.

The Peeling Method

 Example

Write each of the following as a product of prime factors. (a) 110 (b) 7290.

Solution

(a)

```
2|110
5| 55
 | 11
```

∴110=2×5×11

(b)

```
2 |7290
3 |3645
3 |1215
3 | 405
3 | 135
3 | 45
3 | 15
5 | 5
  | 1
```

∴7290=2×3×3×3×3×3×3×5=2×3^6×5

The Factor Tree

 Example

Do the example above using the factor tree method.

Solution

(a)

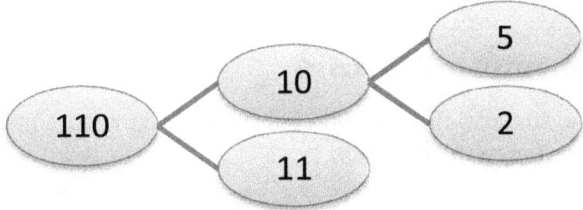

∴ 110 = 2 × 5 × 11

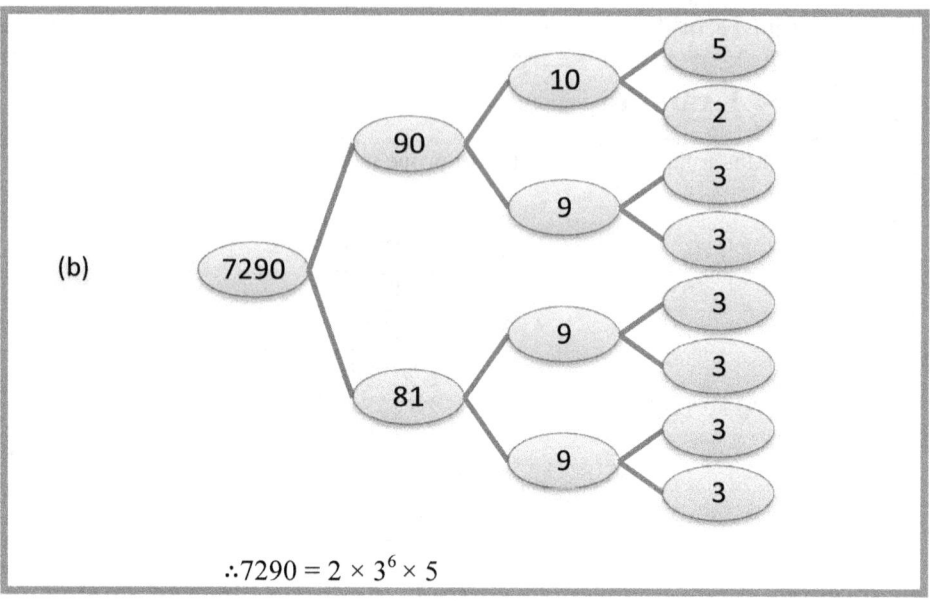

(b)

$$\therefore 7290 = 2 \times 3^6 \times 5$$

Note that the branches of the factor tree end with prime factors.

 Skill Building Exercise 5:1

1. Write the following as a product of prime factors in index form.
    (a) 32   (b) 81   (c) 60   (d) 60   (e) 45
    (f) 63   (g) 51   (h) 48   (i) 243
2. Decompose the following into the product of their prime factors.
    (a) 630   (b) 2200   (c) 2,002   (d) 1728   (e) 5280

5.6 Common Factors

As the name suggest, the common factors of two or more numbers are the factors that are common to the numbers.

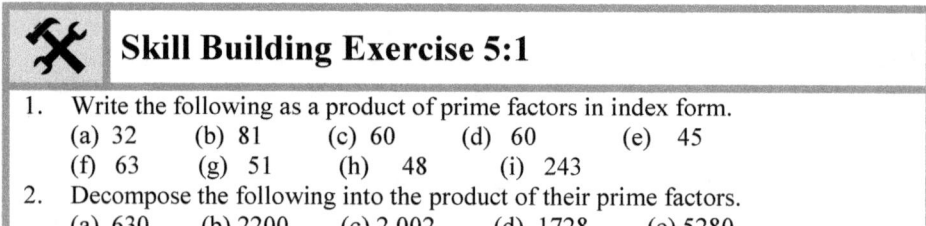 **Example**

List the common factors of 18 and 24.

Solution
Factors of 18 = {1, 2, 3, 6, 9, 18}
Factors of 24 = {1, 2, 3, 4, 6, 8, 12, 24}
∴ Common factors of 18 and 24 = {1, 2, 3, 6}

The Highest Common Factor, HCF

The highest common factor (HCF) of two or more numbers sometimes referred to, as the greatest common divisor (GCD) is the highest or greatest of their common factors.

 Example

Find the HCF of 42 and 48.

Solution
Factors of 42 = {1, 2, 3, 6, 7, 21, 42}
Factors of 48 = {1, 2, 3, 4, 6, 8, 12, 24, 48}
Common factors of 18 and 24 = {1, 2, 3, 6}
∴ Highest common factor (HCF) of 18 and 24 = 6

5.7 Common Multiples

Again, the name suggests that the common multiples of two or more numbers are the multiples of these numbers that are common to the numbers.

 Example

List the first four common multiples of 2 and 3.

Solution
Let M_2 = multiples of 2 and M_3 = multiples of 3
Then, M_2 = {2,4,6,8,10,12,14,16,18,20,22,24,26, ...}
M_3 = {3,6,9,12,15,18,21,24,27, ...}
∴ First four common multiples of 2 and 3 = {6,12,18,24}.

The Least Common Multiple

The Least common multiple (LCM) of two or more members is the smallest of their common multiples. The LCM of the denominations of fractions is called to as the least common denominator (LCD).

Example

Find the LCM of 3 and 4

Solution

Multiples of 3 = {3, 6, 9, 12, 15, 18, 21, 24, 27...}
Multiples of 4 = {4, 8, 12, 16, 20, 24, 28...}
Common multiples of 3 and 4 = {12, 24,...}
∴ LCM of 3 and 4 = 12

The method of listing factors as in the above examples is called the **roster method** is good for finding the HCF and LCM of simple cases. Cases that are more difficult require some prime factorization methods such as the peeling method, the product of prime factors method or the Venn diagram method.

HCF and LCM by Prime Factorization

Whichever prime factorization method is used;

1. *The HCF is the product of all the common prime factors of the numbers.*

2. *The LCM is the product of all the uncommon prime factors and the HCF.*

> HCF = Product of the common factors.
>
> LCM = HCF × Uncommon prime factors.

 Example

1. Find the HCF and LCM of 28 and 42.

    Solution by the Peeling Method

    Simultaneously peel all the numbers as shown below. The HCF requires only partial peeling i.e. peeling that ends when there are no common factors. The LCM requires complete peeling i.e. peeling that ends with the factor 1.

2	28	42
7	14	21
	2	3

    ∴ HCF = 2×7 = 14

2	28	42
7	14	21
2	2	3
3	1	3
	1	1

    LCM = 2 × 7 × 2 × 3 = 84

Solution by product of Prime Factors Method

Write each of the numbers as a product of their primes. Map the common factors as shown. The uncommon prime factors are unmapped.

$$28 = 2 \times 2 \times 7$$
$$\updownarrow \qquad \updownarrow$$
$$42 = 2 \times 3 \times 7$$

HCF $= 2 \times 7 = 14$

LCM = HCF $\times$ uncommon prime factors $= 14 \times 2 \times 3 = 84$

2. Find the HCF and LCM of 15 and 20.

Solution

To acquaint the reader on the application of the both methods above the methods are again used. In an examination use only one method at each instance.

Solution by the peeling method

Peeling is often and better done only up to the point where no two numbers have common factors.

5	15	20
	3	4

In this modified and simplified peeling method,

HCF = product of left column numbers = 5.

LCM = product of left column and bottom row numbers = $5 \times 3 \times 4 = 60$.

By product of prime Factors Method

$15 = 3 \times 5$ and $20 = 2 \times 2 \times 5$

$\therefore$ HCF = 5 and LCM = $2 \times 2 \times 3 \times 5 = 60$.

3. Find the LCM and HCF of 8,12 and 20

Solution

By the peeling method

2	8	12	20
2	4	6	10
	2	3	5

2	8	12	20
2	4	6	10
2	2	3	5
3	1	3	5
5	1	1	5
	1	1	1

HCF = $2 \times 2 = 4$ and LCM = $2 \times 2 \times 2 \times 3 \times 5 = 120$

By using the product of primes method,

$8 = 2 \times 2 \times 2, \quad 12 = 2 \times 2 \times 3, \quad 20 = 2 \times 2 \times 5$

HCF= product of common prime factors $= 2 \times 2 = 4$

LCM=HCF $\times$ uncommon prime factors $= 4 \times 2 \times 3 \times 5 = 120$

Skill Building Exercise 5:2

1. Find the HCF and LCM of:
   (a) 12 and 18   (b) 18 and 16   (c) 96 and 72
   (d) 24 and 21   (e) 12 and 8   (f) 4, 6 and 8
   (g) 18, 24 and 36   (h) 15, 21 and 105 (i)   15, 25 and 75
   (j) 216, 288 and 360
2. Find the HCF and LCM of 15, 21, 30 and 42.

5.8   Squares and Square Roots

To square a number means to multiply the number by itself. This product is called the **perfect square** of the number. Thus, in the field of $\mathbb{N}$, 25 is the square 5 because, $5^2 = 5 \times 5 = 25$.

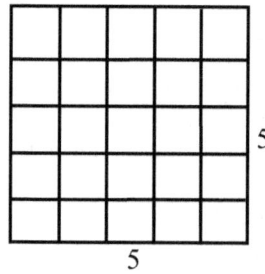

49 is the square of 7 because $49 = 7^2$.

On the other hand 5 is the square root 25 and 7 is the square root 49. The symbol for square root is $\sqrt{}$. Thus,

$$\sqrt{25} = 5 \Leftrightarrow 5^2 = 25 , \; \sqrt{49} = 7 \Leftrightarrow 7^2 = 49$$

Finding the Square Root by Prime Factorization

To find the square root of a number by prime factorization
   (i)   Peel the given number completely using prime factors.
   (ii)   Pair up the repeated prime factors
   (iii) Select one factor from each pair
   (iv)   Multiply the selected factors together.

Example

Find   (a) $\sqrt{64}$ (b) $\sqrt{900}$ (c) $\sqrt{196}$

Solution

(a)   (b)

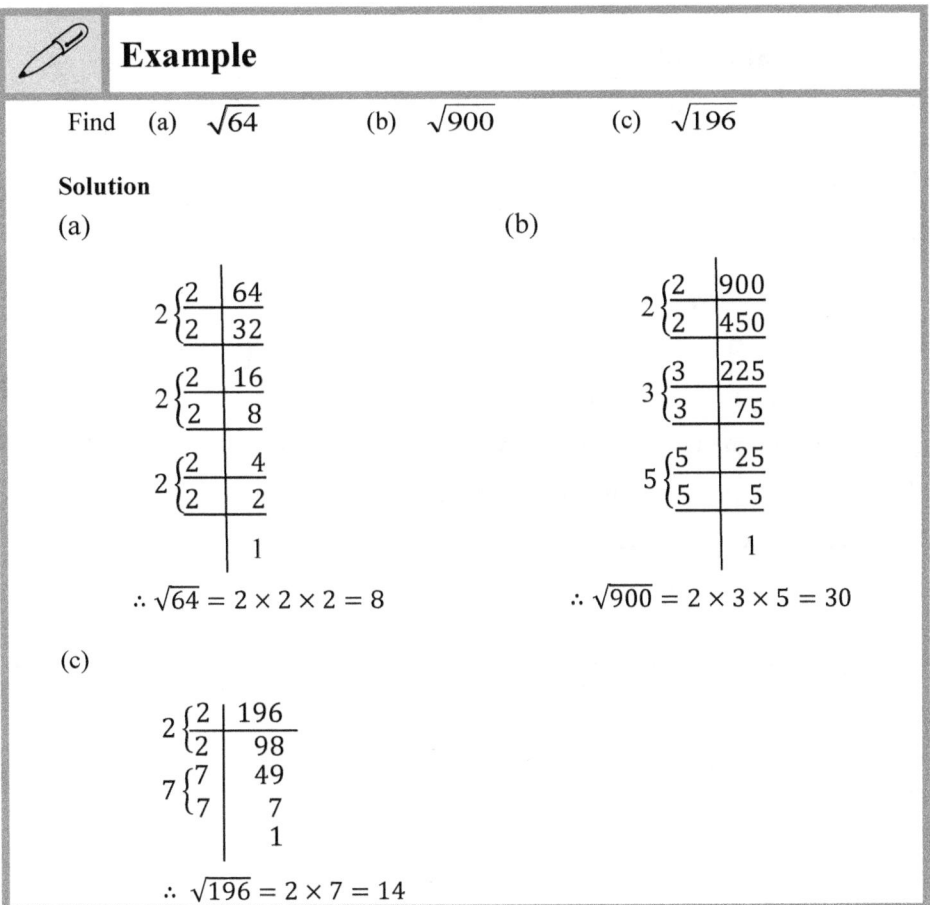

$\therefore \sqrt{64} = 2 \times 2 \times 2 = 8$ $\therefore \sqrt{900} = 2 \times 3 \times 5 = 30$

(c)

$\therefore \sqrt{196} = 2 \times 7 = 14$

5.9   Cubes and Cube Roots

To cube a number means to multiply the number by itself three times. This product is called the **cube** of the number.

9^3 is read as '9 cube' or '9 to the 3rd power'

If a number is a cube of a second number, the second number is its cube root.

$2 \times 2 \times 2 = 8$ means 2 is the cube root of 8. The cube root of 8 is denoted by $\sqrt[3]{8}$.

Example

Evaluate (i) 7^3 (ii) 10^3

Solution

(i) $7^3 = 7 \times 7 \times 7 = 343$ (ii) $10^3 = 10 \times 10 \times 10 = 1000$

Finding the Cube Root by Prime Factorization

To find the cube root of a number,

(a) Peel the given number completely using prime factors

(b) Group the repeated prime factors into three's

(c) Select on factor from each group of three

(d) Multiply together the selected factors.

Example

Find the cube root of (a) 729   (b) 1000

Solution

(a)

$$3\begin{cases}3\\3\\3\end{cases}\begin{array}{c|c} 3 & 729 \\ 3 & 243 \\ 3 & 81 \end{array}$$

$$3\begin{cases}3\\3\\3\end{cases}\begin{array}{c|c} 3 & 27 \\ 3 & 9 \\ 3 & 3 \\ & 1 \end{array}$$

$\therefore \quad \sqrt[3]{729} = 3 \times 3 = 9$

(b)

$$2\begin{cases}2\\2\\2\end{cases}\begin{array}{c|c} 2 & 1000 \\ 2 & 500 \\ 2 & 250 \end{array}$$

$$5\begin{cases}5\\5\\5\end{cases}\begin{array}{c|c} 5 & 125 \\ 5 & 25 \\ 5 & 5 \\ & 1 \end{array}$$

$\therefore \quad \sqrt[3]{1000} = 2 \times 5 = 10$

 **Skill Building Exercise 5:3**

1. Square each of the following
   (a) 13   (b) 31   (c) 17   (d) 20
2. Evaluate each of the following
   (a) 18^2 (b) 28^2 (c) 19^2 (d) 32^2
3. Find the square roots of each of the following numbers
   (a) 784   (b) 289   (c) 6400   (d) 4624
4. Evaluate each of the following
   (a) $\sqrt{625}$ (b) $\sqrt{529}$ (c) $\sqrt{1024}$ (d) $\sqrt{2809}$
5. Cube the following
   (a) 4   (b) 7   (c) 13   (d) 20
6. Evaluate each of the following
   (a) 5^3 (b) 11^3 (c) 8^3 (d) 10^3
7. What is the smallest number by which $2^2 \times 3$ can be multiplied to give a perfect square?

5.10 Divisibility

Divisibility Tests

Divisibility tests are very useful tools for finding the factors of numbers.

 Investigative Activity

Below is a multiplication table for the numbers 2, 3,4,5,6,8,9,10,25 and 50.

×	51	52	53	54	55	56	57	58	59	60
2	102	104	106	108	110	112	114	116	118	120
3	153									
4	204									
5	255									
6	306									
8	408									
9	459									
10	510									
25	1275									
50	2550									3000
100	5100									

1. Complete the table. You may use a calculator.
2. What is common about all the numbers in row 2 (Multiplication table for 2)? What is the general name given to these types of numbers?
3. Add the digits of each of the numbers in row 3(Multiplication table for 3). Is the sum of the digits of each number exactly divisible by 3?
4. Add the digits of each of the numbers in row 8 (Multiplication table for 9). Is the sum of the digits of each number exactly divisible by 9?
5. Are the numbers in row 6 (Multiplication table for 6) all divisible by both 2 and 3?
6. Is the number formed by the last two digits in row 4(Multiplication table for 4) divisible by 4?
7. Is the number formed by the last three digits in row 7(Multiplication table for 8) divisible by 4?
8. What is the last digit of each of the numbers in row 9 (Multiplication table for 10)?
9. Do the numbers in row ten (Multiplication table for 25), end either in 25, 50, 75 or 00?
10. Do the numbers in row eleven (Multiplication table for 50), end either in 50 or 00?

Divisibility Rules

A number is divisible by;

- **2** if its last digit is 0, 2, 4,6, or 8. For instance 64744635<u>8</u>, is divisible by 2.
- **4** if the number determined by the last two digits is divisible by 4. For instance, 2356<u>56</u> is divisible by 4 since 56 is divisible by 4.
- **8** if the number determined by the last three digits is divisible by 8. For instance, 276<u>248</u> is divisible by 8 since 248 is divisible by 8.
- **3** if the sum of its digits is divisible by 3. For instance 87146328 is divisible by 3 since 8+7+1+4+6+3+2+8 =39 and 39 is divisible by 3.
- **9** if the sum of its digits is divisible by 9. For instance 6465357 is divisible by 9 since $6 + 4 + 6 + 5 + 3 + 5 + 7 = 36$ and 36 is divisible by 9.
- **6** if its last digit is 0, 2, 4, 6, 8 and the sum of its digits is divisible by 3. For instance 871463<u>2</u><u>8</u>, is divisible by 6 since it last digit is 8 and $8 + 7 + 1 + 4 + 6 + 3 + 2 + 8 = 39$. 39, is divisible by 3.
- **5** if its last digit is 5 or 0.
- **50** if its last two digits are 50 or 00.
- **25** if its last two digits are 25, 50, 75 or 00.
- **10** or a power of ten if the number of zeros in the number is at least equal to the number of zeros in the power of 10. for instance 70, 700, 7000 are divisible by 10, 100, and 1000 respectively since their last, last two and last three digits are, 0, 00, and 000 respectively.

Skill Building Exercise 5:4

1. Say giving reasons which of the following decimal numbers is divisible by
   (a) 2   (b) 3   (c) 4   (d) 5   (e) 9   (f) 10
      64665, 3689, 85564, 21342, 97965, 76445, 4378, 23490
2. Copy and complete the following table by marking X where applicable.

Number	Divisible by										
	2	3	4	5	6	8	9	10	25	50	100
4750											
9275											
1425											
7200											
5427											

Integration Activity

1. You are paid to plant 6150 fruit trees in an orchard so that each row has exactly the same number of trees. In which of the following ways is it possible to plant them?
   (a) twos (b) threes (c) fours (d) fives (e) eights (f) nines (g) tens.
   Give reasons for your answer in each case.
2. A company produces radios of dimensions 3 cm by 4 cm by 6 cm. You are the container designer of the company. What is the least number of radios that can be packed into the smallest container without any waste of space?
3. In Cameroon presidential elections are normally organized after every 7 years while legislative elections are normally organized after every 5 years. After how many years will the year of presidential and legislative elections always coincide?

Multiple Choice Exercise 5

1. The number of prime numbers between 1 and 20 is:
   [A] 9   [B] 8   [C] 7   [D] 6

2. The first four prime numbers are:
   [A] 1, 2, 3, 4   [B] 1, 3, 5, 7   [C] 2, 3, 5, 7   [D] 2, 4, 6, 8

3. 84 as a product of prime factors is:
   [A] $2^2 \times 3 \times 7$ [B] $2^2 \times 3^3 \times 7$ [C] $2^2 \times 3^2 \times 7^2$ [D] $2^2 \times 3^2 \times 7$

4. As a product of prime factors 200 can be written as:
   [A] 2×5 [B] $2^3 \times 5^2$ [C] $2^2 \times 5^3$ [D] $2^2 \times 5^2$

5. Leaving the answer as the product of prime factors in indexform72 equals:
   [A] $2^2 \times 3^2$ [B] $2^2 \times 3^3$ [C] $2^3 \times 3^2$ [D] $2^3 \times 3^3$

6. The number which is not a prime number is:
   [A] 2   [B] 3   [C] 7   [D] 9

7. The number of factors of a prime number is:
   [A] 3   [B] 2   [C] 1   [D] 0

8. The number which is a prime number is:
   [A] 1   [B] 9   [C] 13   [D] 15

9. 17 is a prime number because:
   [A] it is not divisible by 2   [B] it has only two factors
   [C] it has no factor other than itself   [D] it is a sieve of Erasthodene

10. The number which is a prime number is:
    [A] 57   [B] 61   [C] 63   [D] 69

11. Two of the numbers 11, 21, 31, 77, 112, are prime numbers. The number lying exactly half way between them is:
    [A] 54   [B] 26   [C] 21   [D] 16

12. Three of these numbers 11, 21, 31, 77, 112, have a common factor. The common factor is:
    [A] 7   [B] 11   [C] 14   [D] 2

13. The number is the product of two consecutive prime numbers:
    [A] 21   [B] 18   [C] 8   [D] 15

14. The LCM of 6 and 14 is:
    [A] 14   [B] 24   [C] 42   [D] 84

15. The LCM of 8, 9 and 12 is:
    [A] 29   [B] 72   [C] 96   [D] 108

16. The result of dividing the LCM of 8 and 12 by 3 is:
    [A] 12   [B] 24   [C] 4   [D] 8

17. The HCF of the numbers 30,120 and 125 is:
    [A] 3   [B] 5   [C] 10   [D] 15

18. The HCF of 18, 24 and 36 expressed as a product of prime factors is:
    [A] 2×3 [B] $2 \times 2 \times 3$ [C] $2 \times 3 \times 3$ [D] $2 \times 2 \times 3 \times 3$

19. Dividing the LCM of 24 and 30 by their HCF gives:
    [A] 2   [B] 20   [C] 24   [D] 25

20. The result of dividing the LCM of 12, 16 and 24 by their HCF is:
    [A] 12   [B] 11   [C] 10   [D] 9

21. The result of squaring the number 6 is:

[A] 12   [B] 26   [C] 36   [D] 62

22. The first number is divisible by the second in:

[A] 79 by 9   [B] 11 by 2   [C] 49 by 8   [D] 30 by 6

23. The composite number is:

[A] 41   [B] 81   [C] 47   [D] 31

24. The group of numbers which are all divisible by 5 is:

[A] 124, 333, 315, 266, 391   [B] 135, 211, 330, 274, 252

[C] 135, 205, 330, 275, 365   [D] 232, 250, 365, 225, 210

25. The number 31,186 is:

[A] divisible by 5, but not by 2 or 10. [B] divisible by 2, but not by 5 or 10.

[C] divisible by 2, 5, and 10.   [D] divisible by 5 and 10 but not by 2.

26. The digit that can replace the question sign to make ?3,520 divisible by 9 is:

[A] 5   [B] 7   [C] 6   [D] 8

27. The group of numbers which are divisible by 9 is:

[A] 9,180 and 4,932   [B] 9,180 and 7,625

[C] 2,969 and 4,932   [D] 2,969, 9,180, and 7,625

28. 75,243 for divisibility by 2, 3, 5, 9, or 10.

[A] 5   [B] 3   [C] 2   [D] none of the above

29. The digit(s) which can replace the question sign to make ?3,71 divisible by 3 is/are:

[A] 3 only   [B] 7 only   [C] 1 only   [D] 1, 4, and 7

30. The list of factors of 48 is:

[A] 1, 2, 3, 7, 8, 12, 48   [B] 1, 2, 3, 4, 6, 8, 12, 16, 24, 48

[C] 2, 3, 4, 6, 8, 16, 24   [D] 2, 3, 4, 6, 8, 9, 12, 16, 24

31. The list of factors of 40 is:

[A] 1, 2, 3, 5, 10, 15, 50   [B] 1, 2, 5, 10, 25, 50

[C] 1, 2, 4, 5, 8, 10, 20, 40   [D] 2, 3, 4, 10, 20, 30, 40

32. The prime factorization of 168 is:

[A] $2^3 \times 3 \times 7$ [C] $2^3 \times 3^3 \times 7$ [B] $2^4 \times 3 \times 7$ [D] $2^4 \times 3^3 \times 13$

33. The prime factorization of 540 is:

[A] $2^2 \times 3^3 \times 10$ [B] $2^2 \times 3^3 \times 5$ [C] $2 \times 3^3 \times 5^3$ [D] $2^2 \times 3^4 \times 5$

34. Three different two-digit numbers end in 4. Each is less than 50. The HCF of the numbers is:

[A] 1   [B] 2   [C] 4   [D] 8

Topic 6

RATIONAL NUMBERS AND FRACTIONS

Objectives

At the end of this topic, the learner should be able to:

1. Recognize and identify rational numbers and denote this set.
2. Distinguish between the set of rational numbers and the set of integers and say which one is part of the other.
3. Explain the meaning of a fraction and identify the numerator and denominator of a vulgar fraction.
4. Read and write fractions in the form $\frac{a}{b}$ not a/b.
5. Identify and distinguish between the various types of fractions.
6. Give real examples of fractions and explain their meaning.
7. Identify proper fractions, improper fractions and mixed numbers.
8. State without hesitation the values of fractions with numerator 0 or denominator 0.
9. Identify equivalent fractions.
10. Simplify fractions to their lowest terms.
11. Change improper fractions to mixed numbers and vice versa.
12. Compare fractions using < and >, by expressing them as fractions with a common denominator.
13. Add, subtract, multiply and divide fractions.
14. Express one quantity as a fraction of another.
15. Apply fractions to real life situations.

6.1 Notion of Rational Numbers and Fractions

> **?** | **Brainstorming Exercise**
>
> 1. Suppose you slice a watermelon into two equal parts and eat one slice.
>
>
>
> 2. What quantity of the watermelon have you eaten?
> 3. Can you represent the quantity you have eaten using natural numbers or integers?

The quantity of the watermelon that you have eaten is one out of two slices or half written as $\frac{1}{2}$ which means 'one part out of two parts'. This number is neither a natural number nor an integer. The number is the quotient of two integers. Therefore, there is a need for us to extend the set of integers to a set which will include numbers such as $\frac{1}{2}$. The set of numbers is called the set of rational numbers.

> A **rational number** is a number which can be expressed as a quotient of two integers. The set of rational numbers is denoted by $\mathbb{Q}$.

Examples of rational numbers are $\frac{3}{4}, -\frac{5}{8}, \frac{7}{2}, 1, 0, -6, 1\frac{3}{8}$.

We can represent rational numbers on a number line just like integers.

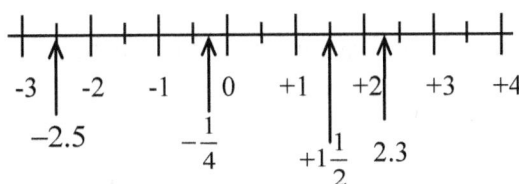

> A **fraction** is a rational number that is not a whole number.

Fractions arise when one quantity is divided by another and there is a remainder. Examples of fractions are $\frac{2}{3}, \frac{5}{4}, 1\frac{1}{4}$. A fraction such as $\frac{2}{3}$ means, 'divide 2 things by 3' or 'divide a quantity into 3 parts and take 2 parts'. In fractions such as $\frac{2}{3}$ and $\frac{5}{4}$, the top number is called the **numerator** or **dividend**, while the bottom number is called the **denominator** or **divisor**.

$$\frac{2}{3} \begin{array}{l} \leftarrow \text{ numerator or dividend} \\ \leftarrow \text{ denominator or divisor} \end{array}$$

Examples of Fractions in Real life

1. The rectangle below represents a loaf of bread. You are asked to slice the loaf into three equal slices and eat two slices.
   (a) Show how you will slice it and shade the part you will eat.
   (b) What quantity will you eat?

2. (a) Explain how you will share five oranges equally to four children.
   (b) What quantity will one child take?
3. What fraction of the following shapes, are circular?

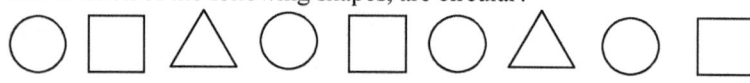

4. What quantity of the circular flower bed is white?

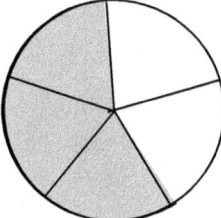

In the examples above we see that:
1. A possible way of slicing the loaf is as shown and the quantity you will eat is $\frac{2}{3}$ (shaded). The quantity left is $\frac{1}{3}$ (not shaded)

2. To share five oranges equally to four children one child will take one orange each and still partake in one quarter of a full orange.

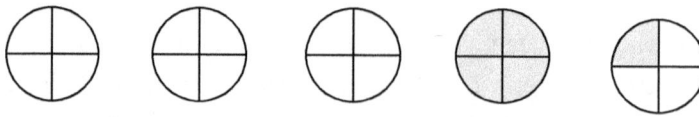

If one child decides to slice his full orange into 4 parts too, then he will have five $\frac{1}{4}$ oranges. This shows that $\frac{5}{4} = 1\frac{1}{4}$.

3. There are 9 shapes and 4 are circular, so $\frac{4}{9}$ are circular.
4. The fraction which is white is $\frac{2}{5}$.

6.2 Types of Fractions

Fractions such as $\frac{1}{5}, \frac{3}{4}, \frac{2}{3}$ etc. whose numerators are less than their denominators are called **proper fractions**.

For a proper fraction, Numerator $<$ *Denominator*

Fractions such as $\frac{11}{5}, \frac{9}{4}, \frac{7}{3}$ etc whose numerators are greater than their denominators are called **improper fractions**.

For an improper fraction, Numerator $>$ *Denominator*

From above, it was established that $\frac{5}{4} = 1\frac{1}{4}$. $1\frac{1}{4}$ is read, one and one quarter and is an example of a **mixed number** or **mixed fraction**.

Mixed fraction = whole number + proper fraction

 Exercise 6:1

1. State the numerator of the following fractions.
    (a) $\frac{2}{3}$ (b) $\frac{5}{8}$ (c) $\frac{7}{4}$
2. State the denominator of the following fractions.
    (a) $\frac{2}{3}$ (b) $\frac{5}{8}$ (c) $\frac{7}{4}$
3. Read aloud the following fractions.
    (a) $\frac{4}{3}$ (b) $1\frac{5}{8}$ (c) $\frac{7}{10}$ (d) $\frac{2}{3}$ (e) $\frac{9}{5}$ (f) $5\frac{2}{3}$
4. Write down the following fractions.
    (a) two-thirds   (b) three and four-fifths   (c) sixteen over three
    (d) five-ninths   (e) four-thirds   (f) eight and one-tenth
5. Classify the following fractions as proper, improper or mixed fractions.
    (a) $\frac{4}{3}$ (b) $1\frac{5}{8}$ (c) $\frac{7}{10}$ (d) $\frac{2}{3}$ (e) $\frac{9}{5}$ (f) $5\frac{2}{3}$

Fractions with zero Numerators or Denominators

 Investigative Activity

1. Use your calculator to compute the following.
    (a) $\frac{0}{7}$ (b) $\frac{7}{0}$ (c) $\frac{0}{0}$ (d) $\frac{0}{15}$ (e) $\frac{30}{0}$ (f) $\frac{0}{100}$
2. What conclusion do you draw?

Any fraction such as $\frac{0}{7}$ with numerator zero and a denominator other than zero is equal to 0. On the other hand any fraction such as $\frac{0}{0}$ and $\frac{7}{0}$ with denominator zero has no meaning.

6.3   Equivalent Fractions

Investigative Activity

(1) In each of the following rectangles, write down the fraction which is shaded in terms of the number of parts.

(a)

(b)

(c)

(d)

(2) What conclusion do you draw?

In (a), (b), (c) and (d) the fractions that are shaded in terms of the number of parts are $\frac{1}{2}, \frac{2}{4}, \frac{4}{8}, \frac{8}{16}$ respectively.

Since virtually the same quantity has been shaded in each case, it means
$$\frac{1}{2} = \frac{2}{4} = \frac{4}{8} = \frac{8}{16}.$$

Fractions such as $\frac{1}{2}, \frac{2}{4}, \frac{4}{8}$ and $\frac{8}{16}$ which represent the same quantities are called **equivalent fractions**.

By multiplying $\frac{1}{2}$ in turns by $\frac{2}{2}, \frac{4}{4}$, and $\frac{8}{8}$, we can see that:
$$\frac{2}{4} = \frac{1}{2} \times \frac{2}{2}, \frac{4}{8} = \frac{1}{2} \times \frac{4}{4}, \frac{8}{16} = \frac{1}{2} \times \frac{8}{8}.$$

Therefore, the value of a fraction remains the same if both numerator and denominator are multiplied by the same quantity.

82

When two fractions such as $\frac{1}{2}$ and $\frac{8}{16}$ are equivalent, we write $\frac{1}{2} \equiv \frac{8}{16}$ in the strictest sense though $\frac{1}{2} = \frac{8}{16}$ is permissible.

 Example

What number should replace the question sign in $\frac{2}{3} = \frac{?}{15}$?

By dividing 15 by 3 we have 5. Therefore the denominator of the LHS has been multiplied by 5 to give the denominator of the RHS. Hence the numerator should equally be multiplied by 5.

$$\therefore \frac{2}{3} \equiv \frac{2}{3} \times \frac{5}{5} \equiv \frac{10}{15}$$

Simplifying Fractions to their Lowest Terms

A fraction is said to be in its **lowest terms**, if the numerator and denominator have no common prime factors. Examples of fractions in their lowest terms are $\frac{2}{3}, \frac{4}{9}, \frac{8}{15}$ etc.

> To simplify a fraction to its lowest terms divide both numerator and denominator by their HCF.

 Example

Simplify the following: (a) $\frac{28}{36}$ (b) $\frac{15}{30}$

Solutions

(a) HCF of 28 and 36 is 4

$$\frac{\overset{7}{\cancel{28}}}{\underset{9}{\cancel{36}}} = \frac{7}{9}$$

(b) HCF of 15 and 30 is 15.

$$\frac{\overset{1}{\cancel{15}}}{\underset{2}{\cancel{30}}} = \frac{1}{2}$$

Real life Example

840 out of the 3960 students who sat for an examination failed. What fraction of the students in its lowest terms passed?

Number who passed = Number who sat - Number who failed
$$= 3960 - 840 = 3120$$

Therefore, fraction who passed $= \frac{3120}{3960} = \frac{26}{33}$

Skill Building Exercise 6:1

1. Write down the missing number in each of the following cases.

(a) $\frac{3}{5} = \frac{}{100}$ (b) $\frac{3}{2} = \frac{}{24}$ (c) $\frac{5}{8} = \frac{}{40}$ (d) $\frac{2}{3} = \frac{24}{}$ (e) $\frac{7}{3} = \frac{28}{}$ (f) $\frac{4}{9} = \frac{60}{}$

2. Simplify.

(a) $\frac{33}{120}$ (b) $\frac{72}{48}$ (c) $\frac{64}{80}$ (d) $\frac{24}{18}$ (e) $\frac{28}{21}$ (f) $\frac{60}{81}$

6.4 Converting Mixed Numbers to Improper Fractions and Vice Versa

1. To convert a mixed number to an improper fraction, multiply the denominator of the fractional part by the whole number and add the result to the numerator of the fractional part to get the numerator of the improper fraction. The denominator of the fractional part is still the denominator of the mixed number.

2. To convert an improper fraction to a mixed number, divide the numerator by the denominator write the result as the whole number part and the fractional part where the fractional part is the quotient of the remainder and the denominator of the improper fraction.

Example

1. Convert the following mixed numbers to improper fractions.

(a) $1\frac{4}{5}$ (b) $5\frac{7}{8}$

Solutions

(a) $1\frac{4}{5} = \frac{5 \times 1 + 4}{5} = \frac{9}{5}$

(b) $5\frac{7}{8} = \frac{8 \times 5 + 7}{8} = \frac{47}{8}$

2. Convert the following improper fractions to mixed numbers.

(a) $\frac{37}{8}$

(b) $\frac{18}{5}$

Solutions

(a) $\frac{37}{8} = \frac{32+5}{8} = \frac{4(8)}{8} + \frac{5}{8} = 4 + \frac{5}{8} = 4\frac{5}{8}$

(b) $\frac{18}{5} = \frac{15+3}{5} = 3 + \frac{3}{5} = 3\frac{3}{5}$

3. Convert the following mixed numbers to improper fractions.

(a) $1\frac{4}{5}$

(b) $5\frac{7}{8}$

Solutions

(a) $1\frac{4}{5} = \frac{5 \times 1 + 4}{5} = \frac{9}{5}$

(b) $5\frac{7}{8} = \frac{8 \times 5 + 7}{8} = \frac{47}{8}$

 Skill Building Exercise 6:2

1. Convert the following mixed numbers into improper fractions

(a) $2\frac{2}{3}$ (b) $5\frac{1}{5}$ (c) $1\frac{3}{4}$ (d) $3\frac{1}{5}$ (e) $2\frac{3}{5}$

2. Convert the following improper fractions into mixed numbers

(a) $\frac{11}{3}$ (b) $\frac{15}{2}$ (c) $\frac{9}{4}$ (d) $\frac{11}{7}$ (e) $\frac{13}{3}$

6.5 Comparing and Ordering Fractions

 Discussion Exercise

1. What can you say about the denominators of the fractions $\frac{2}{7}$ and $\frac{3}{7}$?

2. What can you say about the numerators of the fractions above?

3. You are asked to compare the fractions. What do you do?

4. How do you make fractions with different denominators have the same denominators?

5. Explain what you will do to compare the fractions $\frac{2}{3}$ and $\frac{4}{5}$.

6. How will you compare $1\frac{4}{7}$ and $1\frac{9}{14}$?

(i) To compare fractions with the same denominators, compare the numerators.

(ii) To compare fractions with different denominators, first convert the fractions to equivalent fractions with the same denominators. Here it is advisable to make the LCM their denominators.

(iii) To compare mixed numbers first convert them to improper fractions the convert the fractions to equivalent fractions with the same denominators.

 ## Examples

Compare the following using $>$, $<$ or $=$

(a) $\frac{5}{8}$ and $\frac{3}{8}$ (b) $\frac{7}{10}$ and $\frac{9}{10}$ (c) $\frac{3}{4}$ and $\frac{7}{10}$ (d) $\frac{1}{6}$ and $\frac{3}{8}$ (e) $\frac{2}{3}$ and $\frac{4}{5}$

Solutions

(a) $\frac{5}{8} > \frac{3}{8}$, since $5 > 3$.

(b) $\frac{7}{10} < \frac{9}{10}$, since $7 < 9$.

(c) LCM of the denominators is 20.

$$\frac{3}{4} = \frac{3\times5}{4\times5} = \frac{15}{20} \text{ and } \frac{7}{10} = \frac{7\times2}{10\times2} = \frac{14}{20}$$

Since $15 > 14$, $\dfrac{3}{4} > \dfrac{7}{10}$

(d) LCM of the denominators is 24.

$$\frac{1}{6} = \frac{1\times4}{6\times4} = \frac{4}{24} \text{ and } \frac{3}{8} = \frac{3\times3}{8\times3} = \frac{9}{24}$$

Since $4 < 9$, $\dfrac{1}{6} < \dfrac{3}{8}$

(e) The LCM of the denominators is 15. So $\frac{2}{3} = \frac{2\times5}{3\times5} = \frac{10}{15}$ and $\frac{4}{5} = \frac{4\times3}{5\times3} = \frac{12}{15}$

Since $12 > 10$, $\frac{2}{3} < \frac{4}{5}$.

 ## Real life Example

At the same time Yuh and Bih cover distances of $1\frac{4}{9}$ km and $1\frac{5}{12}$ km. Who runs faster?

The LCM of the denominators 9 and 12 is 36.

$\Rightarrow$ Yuh runs $1\frac{4}{9} = \frac{13}{9} = \frac{13\times4}{9\times4} = \frac{52}{36}$ and Bih runs $1\frac{5}{12} = \frac{17}{12} = \frac{17\times3}{12\times3} = \frac{51}{36}$.

Since $52 > 51$, it means Yuh runs faster than Bih.

Skill Building Exercise 6:3

1. Compare the following using the symbol $<$, $>$ or $\equiv$

   (a) $\dfrac{7}{9}$ $\dfrac{6}{9}$ (b) $\dfrac{2}{3}$ $\dfrac{4}{5}$ (c) $\dfrac{3}{6}$ $\dfrac{2}{4}$ (d) $\dfrac{3}{4}$ $\dfrac{7}{9}$

2. Arrange the following in ascending order of magnitude.

   $\dfrac{2}{3}, \dfrac{1}{2}, \dfrac{7}{8}, \dfrac{1}{4}, \dfrac{3}{4}, \dfrac{1}{3}, \dfrac{2}{5}, \dfrac{4}{5}, \dfrac{8}{9}, \dfrac{6}{7}$

3. Arrange each of the following groups of fractions in order of increasing magnitude.

   (a) $\dfrac{3}{6}, \dfrac{5}{6}, \dfrac{7}{6}, \dfrac{4}{6}, \dfrac{2}{6}$ (b) $\dfrac{6}{8}, \dfrac{7}{8}, \dfrac{2}{8}, \dfrac{4}{8}$ (c) $\dfrac{9}{7}, \dfrac{9}{2}, \dfrac{9}{5}, \dfrac{9}{11}, \dfrac{9}{8}$ (d) $\dfrac{11}{3}, \dfrac{12}{4}, \dfrac{10}{5}, \dfrac{16}{4}$

4. Compare the following using the symbol $<$, $>$ or $\equiv$

   (a) $\dfrac{17}{9}$ $1\dfrac{6}{9}$ (b) $\dfrac{32}{30}$ $\dfrac{4}{5}$ (c) $\dfrac{33}{66}$ $\dfrac{22}{44}$ (d) $1\dfrac{3}{4}$ $1\dfrac{7}{9}$

5. Arrange the following in descending order of magnitude

   $1\dfrac{2}{3}, 3\dfrac{1}{2}, \dfrac{17}{8}, 3\dfrac{1}{4}, \dfrac{13}{3}, 5\dfrac{1}{3}, 4\dfrac{2}{5}, \dfrac{16}{5}, \dfrac{18}{6}, \dfrac{20}{6}$

6.6 Operations with Fractions

Adding and subtracting Proper and Improper Fractions

 Investigative Activity

1. In the above rectangle state the fraction of the rectangle that
   (a) has the colour
   (b) has the colour

2. What fraction of the rectangle is coloured?
3. Explain how you will use the diagram to evaluate.
   (i) $\dfrac{3}{10} + \dfrac{4}{10}$ (ii) $\dfrac{7}{10} - \dfrac{3}{10}$
4. From the above deduce how you would add fractions with unequal denominators.

(a) To add or subtract fractions with equal denominators, keep the common denominator and add or subtract the numerators.

(b) To add or subtract fractions with unequal denominators, first use the idea of equivalent fractions to convert each of the fractions to equivalent fractions with the same denominator. It is often advisable to make the least common denominator (LCD) their denominators.

 Examples

Evaluate the following.

(a) $\frac{3}{5} + \frac{1}{5}$ (b) $\frac{9}{10} + \frac{2}{10}$ (c) $\frac{3}{5} - \frac{1}{5}$ (d) $\frac{9}{10} - \frac{2}{10}$

(e) $\frac{5}{6} + \frac{3}{4}$ (f) $\frac{2}{3} + \frac{3}{5} + \frac{1}{6}$ (g) $\frac{7}{8} - \frac{1}{4}$ (h) $\frac{5}{6} - \frac{2}{3} - \frac{1}{10}$

Solutions

(a) $\frac{3}{5} + \frac{1}{5} = \frac{4}{5}$ (b) $\frac{9}{10} + \frac{2}{10} = \frac{11}{10}$ (c) $\frac{3}{5} - \frac{1}{5} = \frac{2}{5}$ (d) $\frac{9}{10} - \frac{2}{10} = \frac{7}{10}$

(e) $\frac{5}{6} + \frac{3}{4} = \frac{10}{12} + \frac{9}{12} = \frac{19}{12} = 1\frac{7}{12}$ (f) $\frac{2}{3} + \frac{3}{5} + \frac{1}{6} = \frac{20}{30} + \frac{18}{30} + \frac{5}{30} = \frac{43}{30} = 1\frac{13}{30}$

(g) $\frac{7}{8} - \frac{2}{8} = \frac{5}{8}$ (h) $\frac{5}{6} - \frac{2}{3} - \frac{1}{10} = \frac{25}{30} - \frac{20}{30} - \frac{3}{30} = \frac{2}{30} = \frac{1}{15}$

The following equivalent layout may be used.

(e) $\frac{5}{6} + \frac{3}{4} = \frac{10+9}{12} = \frac{19}{12} = 1\frac{7}{12}$ (f) $\frac{2}{3} + \frac{3}{5} + \frac{1}{6} = \frac{20+18+5}{30} = \frac{43}{30} = 1\frac{13}{30}$

(g) $\frac{7}{8} - \frac{1}{4} = \frac{7-2}{8} = \frac{5}{8}$ (h) $\frac{5}{6} - \frac{2}{3} - \frac{1}{10} = \frac{25-20-3}{30} = \frac{2}{30} = \frac{1}{15}$

 Skill Building Exercise 6:4

1. Evaluate the following
   (a) $\frac{4}{7} + \frac{2}{7}$ (b) $\frac{6}{11} + \frac{4}{11}$ (c) $\frac{7}{8} + \frac{5}{8}$ (d) $\frac{1}{9} + \frac{4}{9}$ (e) $\frac{3}{10} + \frac{1}{10}$

2. Evaluate the following
   (a) $\frac{4}{9} - \frac{2}{9}$ (b) $\frac{6}{7} - \frac{4}{7}$ (c) $\frac{7}{8} - \frac{5}{8}$ (d) $\frac{9}{13} - \frac{4}{13}$ (e) $\frac{3}{5} - \frac{1}{5}$

3. Add the following
   (a) $\frac{2}{3} + \frac{3}{4}$ (b) $\frac{6}{7} + \frac{4}{9}$ (c) $\frac{3}{5} + \frac{5}{8}$ (d) $\frac{1}{6} + \frac{4}{5}$ (e) $\frac{3}{5} + \frac{1}{10}$

4. Compute the following

(a) $\dfrac{2}{3} - \dfrac{2}{9}$ (b) $\dfrac{8}{9} - \dfrac{4}{7}$ (c) $\dfrac{7}{8} - \dfrac{5}{11}$ (d) $\dfrac{9}{13} - \dfrac{3}{7}$ (e) $\dfrac{3}{5} - \dfrac{1}{4}$

5. Evaluate and simplify the following where appropriate.

(a) $\dfrac{2}{5} + \dfrac{2}{3} + \dfrac{3}{4}$ (b) $\dfrac{6}{7} + \dfrac{2}{3} - \dfrac{4}{9}$ (c) $\dfrac{3}{5} + \dfrac{5}{8} + \dfrac{3}{4}$ (d) $\dfrac{1}{3} + \dfrac{1}{6} + \dfrac{4}{5}$ (e) $\dfrac{3}{5} - \dfrac{1}{2} + \dfrac{1}{10}$

Adding and subtracting mixed numbers

 Discussion Exercise

Explain the steps you would take to find the sum of $6\dfrac{3}{5} + 4\dfrac{3}{4}$.

We can compute the whole number part and the fractional part separately. However, it is even better to first convert the mixed numbers to improper fractions before adding or subtracting.

 Examples

Evaluate the following

(a) $1\dfrac{3}{10} + 2\dfrac{1}{10}$ (b) $4\dfrac{5}{6} - 1\dfrac{1}{6}$ (c) $6\dfrac{3}{5} + 4\dfrac{3}{4}$

Solutions

(a) $1\dfrac{3}{10} + 2\dfrac{1}{10} = \dfrac{13}{10} + \dfrac{21}{10} = \dfrac{34}{10} = \dfrac{17}{5} = 3\dfrac{2}{5}$

(b) $4\dfrac{5}{6} - 1\dfrac{1}{6} = \dfrac{29}{6} - \dfrac{7}{6} = \dfrac{22}{6} = \dfrac{11}{3} = 3\dfrac{2}{3}$

(c) $6\dfrac{3}{5} + 4\dfrac{3}{4} = \dfrac{33}{5} + \dfrac{19}{4} = \dfrac{132}{20} + \dfrac{95}{20} = \dfrac{227}{20} = 11\dfrac{7}{20}$

Real life Example

1. A taxi man uses $10\frac{2}{5}$ litres of fuel on Monday and $9\frac{3}{4}$ litres on Tuesday. How many litres altogether does he use for these two days?

    Solutions

    Number of litres used for two days $= 10\frac{2}{5} + 9\frac{3}{4} = \frac{52}{5} + \frac{39}{4} = \frac{208+195}{20}$

    $$= \frac{403}{20} = 20\frac{3}{20}$$

2. $\frac{4}{11}$ of the customers in a restaurant eat achu, $\frac{2}{11}$ eat rice $\frac{3}{16}$ eat beans and the rest eat fufu corn. What fraction eats fufu corn?

    Solutions

    Fraction who eat fufu corn $= 1 - \left(\frac{4}{11} + \frac{2}{11} + \frac{3}{16}\right)$

    $$= 1 - \frac{129}{176} = \frac{176}{176} - \frac{129}{176} = \frac{47}{176}$$

 Skill Building Exercise 6:5

1. Add and simplify the following
    (a) $7\frac{2}{5} + 2\frac{4}{5} + 5\frac{3}{5}$ (b) $3\frac{6}{7} + 4\frac{2}{7} + 2\frac{4}{7}$ (c) $2\frac{3}{5} + 3\frac{5}{8} + 5\frac{3}{4}$

    (d) $6\frac{1}{3} + 2\frac{1}{6} + 3\frac{4}{5}$ (e) $4\frac{3}{5} + 2\frac{1}{2} + 1\frac{1}{10}$

2. Compute and simplify the following
    (a) $9\frac{2}{5} + 3\frac{4}{5} - 5\frac{3}{5}$ (b) $4\frac{6}{7} - 2\frac{2}{7} - 1\frac{4}{7}$ (c) $2\frac{3}{5} - 3\frac{5}{8} + 5\frac{3}{4}$

    (d) $6\frac{1}{3} - 2\frac{1}{6} - 3\frac{4}{5}$ (e) $4\frac{3}{5} - 2\frac{1}{2} + 1\frac{1}{10}$

 Real life Exercise

To create a design, a tailor cut off $2\frac{3}{4}$ cm from a green piece of cloth of length $14\frac{5}{8}$ cm and replaced it with a red strip of length $3\frac{4}{5}$ cm. How long is the final piece of cloth? (Ignore any bends or folds).

Multiplying fractions

A product represents 'how many times a given quantity', or 'how many of a given quantity'. Thus;

The product of $\frac{4}{5}$ and 2 is the same as $\frac{4}{5} \times 2$ or $\frac{4}{5}$ of 2.

The product of $\frac{4}{5}$ and $\frac{2}{3}$ is the same as $\frac{4}{5} \times \frac{2}{3}$ or $\frac{4}{5}$ of $\frac{2}{3}$.

Investigative Activity

Let's investigate how to evaluate $\frac{4}{5} \times \frac{2}{3}$.

(1) Cut out a rectangle containing 15 squares from a square grid paper as shown on the right. What fraction of the rectangle is one of the squares?

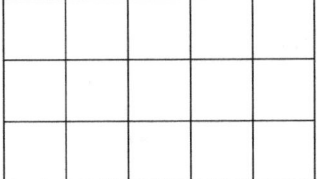

(2) Cut out $\frac{4}{5}$ of the rectangle. You should have a grid which looks like that on the right.

(3) Cut out $\frac{2}{3}$ of this rectangle. You should have a grid which looks like that on the right.

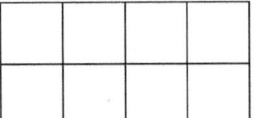

(4) By counting squares, what fraction of the original rectangle is this rectangle?

The rectangle in (3) is $\frac{2}{3}$ of the rectangle in (2) $\left(\text{i. e } \frac{2}{3} \text{ of } \frac{4}{5} \right)$.

By counting squares, we can see that there are 8 squares in the rectangle in (3). So the rectangle in (3) is $\frac{8}{15}$ of the rectangle in (1).

Multiplying the numerators together and the denominators together we see that;

$$\frac{4}{5} \times \frac{2}{3} = \frac{8}{15} \dots \dots \dots \dots \dots ①$$

Hence, it can be deduced that,

To multiply two fractions, multiply the numerators together and the denominators together.

If there are any common factors, these factors should first be cancelled to

simplify the fractions. Thus $\dfrac{\overset{1}{\cancel{2}}}{\underset{1}{\cancel{3}}} \times \dfrac{\overset{1}{\cancel{3}}}{\underset{4}{\cancel{8}}} = \dfrac{1}{4}$.

 Example

Evaluate $\dfrac{18}{35} \times \dfrac{14}{27}$	**Solution** $\dfrac{\overset{2}{\cancel{18}}}{\underset{5}{\cancel{35}}} \times \dfrac{\overset{2}{\cancel{14}}}{\underset{3}{\cancel{27}}} = \dfrac{4}{15}$

 Real life Example

Jane took $\dfrac{2}{3}$ of the money she had in her house to school. At school, she used $\dfrac{3}{4}$ of what she took. What fraction of the money she had in her house did she use?

Solution

$$\frac{3}{4} \text{ of } \frac{2}{3} = \frac{\overset{1}{\cancel{3}}}{\underset{2}{\cancel{4}}} \times \frac{\overset{1}{\cancel{2}}}{\underset{1}{\cancel{3}}} = \frac{1}{2}$$

Multiplying Fractions by Whole Numbers

 Brainstorming Exercise

1. What is the value of $\dfrac{15}{1}$?
2. Explain how you will use your idea in to multiply fractions and whole numbers.

If one of the numbers is a whole number, we can write it as a fraction with denominator 1, and follow the method in ② above. Thus,

$$\frac{3}{5} \times 15 = \frac{3}{5} \times \frac{\overset{3}{15}}{\underset{1}{1}} = 9$$

If the numerator is multiplied by the whole number and the denominator is maintained, the result will be

$$\frac{3}{5} \times 15 = \frac{45}{5} = 9$$

Therefore,

> To multiply a fraction by a whole number, multiply the numerator by the whole number and maintain the denominator.
> If the fractions are mixed numbers, first convert them to improper fractions.

 Examples

Evaluate the following (a) $\frac{3}{11} \times 2$ (b) $5\frac{3}{8} \times 4$ (c) $\frac{3}{4}$ of 300 (d)

Solution

(a) $\frac{3}{11} \times 2 = \frac{3 \times 2}{11} = \frac{6}{11}$

(b) $5\frac{3}{8} \times 4 = \frac{43}{\underset{2}{8}} \times \overset{1}{4} = 21\frac{1}{2}$

(c) $\frac{3}{4}$ of 300 $= \frac{3}{4} \times 300 = 225$

 Real life Example

Mary had 357,000 Francs in the credit union. She signed out $\frac{2}{3}$ of it. How much did she sign out?

Solution

Amount signed out $= \frac{2}{3}$ of 357000 $= \frac{2}{3} \times 357000 = 238000$ Francs

Multiplying Mixed Numbers

> To multiply mixed numbers, first convert the mixed numbers to improper fractions.

Examples

Simplify (i) $2\frac{1}{2} \times 1\frac{1}{4}$ (ii) $4\frac{2}{3} \times 2\frac{3}{5}$.

Solution

(i) $2\frac{1}{2} \times 1\frac{1}{4} = \frac{5}{2} \times \frac{5}{4} = \frac{25}{8} = 3\frac{1}{8}$ (ii) $4\frac{2}{3} \times 2\frac{3}{5} = \frac{14}{3} \times \frac{13}{5} = \frac{182}{15} = 12\frac{2}{15}$.

Real life Example

A stone is $3\frac{1}{2}$ times heavier than a stone of weight $7\frac{1}{2}$ kg. How heavy is the stone?

Solution

The weight of the stone is $3\frac{1}{2} \times 7\frac{1}{2}$ kg $= \frac{7}{2} \times \frac{15}{2} = \frac{105}{4} = 26\frac{1}{4}$ kg.

Skill Building Exercise 6:6

Evaluate the following

1. $\frac{2}{3}$ of 60 2. $\frac{3}{4}$ of 180 3. $\frac{5}{8}$ of $\frac{3}{4}$ 4. $\frac{2}{5}$ of $\frac{75}{80}$

5. $8 \times \frac{3}{4}$ 6. $12 \times \frac{1}{4}$ 7. $\frac{2}{3} \times 9$ 8. $\frac{5}{7} \times 14$

9. $\frac{3}{4} \times \frac{8}{11}$ 10. $\frac{3}{5} \times \frac{2}{9}$ 11. $\frac{5}{6} \times \frac{7}{10}$ 12. $\frac{4}{5} \times \frac{15}{16}$

13. $\frac{1}{5} \times \frac{2}{3} \times \frac{1}{4}$ 14. $\frac{3}{4} \times \frac{1}{6} \times \frac{2}{5}$ 15. $\frac{1}{2} \times \frac{9}{10} \times \frac{2}{3}$ 16. $2\frac{1}{2} \times 4$

17. $1\frac{2}{5} \times 10$ 18. $12 \times 6\frac{1}{2}$ 19. $8 \times 3\frac{3}{4}$ 20. $2\frac{1}{2} \times 1\frac{1}{4}$

21. $3\frac{1}{3} \times 1\frac{3}{4}$ 22. $4\frac{2}{3} \times 2\frac{3}{5}$ 23. $6 \times 5\frac{3}{5} \times 1\frac{2}{3}$ 24. $6\frac{2}{3} \times 7 \times 1\frac{1}{5}$

25. $2\frac{1}{6} \times 1\frac{7}{8} \times 5\frac{1}{3}$ 26. $2\frac{3}{5} \times 1\frac{5}{6} \times 1\frac{1}{2}$

Reciprocals

Investigative Activity

1. Evaluate the following (i) $\frac{3}{4} \times \frac{4}{3}$ (ii) $\frac{5}{7} \times \frac{7}{5}$ (iii) $3 \times \frac{1}{3}$.
2. What conclusion do you draw?

In each case, the product is 1. If two numbers are such that their product is 1, we say one number is the **reciprocal** or the **multiplicative inverse** of the other and vice versa. A number and its reciprocal have the same sign. Thus $-\frac{7}{3}$ is the reciprocal of $-\frac{3}{7}$.

Division of Fractions

Consider $\frac{4}{5} \div \frac{2}{3}$. This can be written as $\frac{\frac{4}{5}}{\frac{2}{3}}$.

A fraction such as $\frac{\frac{4}{5}}{\frac{2}{3}}$ is called a **complex fraction.**

Multiplying numerator and denominator of any fraction by the same quantity does not change the value of the fraction.

$$\frac{\frac{4}{5}}{\frac{2}{3}} = \frac{\frac{4}{5} \times \frac{3}{2}}{\frac{2}{3} \times \frac{3}{2}} = \frac{4}{5} \times \frac{3}{2}.$$

Reciprocals

Therefore,
1. To divide by a fraction, simply multiply by the reciprocal of the fraction.
2. In dividing fractions, if the dividend or the divisor is a mixed number, first convert it to an improper fraction.

 Example

Evaluate (i) $\frac{2}{3} \div \frac{3}{4}$ (ii) $\frac{9}{10} \div \frac{3}{5}$ (iv) $1\frac{4}{5} \div 3$ (v) $5 \div 3\frac{1}{5}$

Solution

(i) $\frac{2}{3} \div \frac{3}{4} = \frac{2}{3} \times \frac{4}{3} = \frac{8}{9}$

(ii) $\frac{9}{10} \div \frac{3}{5} = \frac{\overset{3}{\cancel{9}}}{\underset{2}{\cancel{10}}} \times \frac{\overset{1}{\cancel{5}}}{\cancel{3}} = \frac{3}{2} = 1\frac{1}{2}$

(iii) $1\frac{4}{5} \div 3 = \frac{9}{5} \times \frac{1}{3} = \frac{3}{5}$

(v) $5 \div 3\frac{1}{5} = 5 \div \frac{16}{5} = 5 \times \frac{5}{16} = \frac{25}{16}$

Skill Building Exercise 6:7

1. State the reciprocal of each of the following

    (a) $\dfrac{7}{3}$ (b) $-\dfrac{2}{9}$ (c) $\dfrac{3}{4}$ (d) $\dfrac{5}{8}$ (e) $-\dfrac{11}{6}$

2. Evaluate (a) $9 \div \dfrac{3}{7}$ (b) $\dfrac{9}{10} \div 6$

3. Simplify (a) $\dfrac{1}{10} \div \dfrac{3}{5}$ (b) $\dfrac{3}{4} \div \dfrac{7}{8}$ (c) $\dfrac{\frac{2}{3}}{\frac{6}{9}}$ (d) $\dfrac{\frac{5}{6}}{\frac{1}{2}}$

4. Compute (a) $3\dfrac{1}{2} \div 4\dfrac{1}{2}$ (b) $\dfrac{5}{6} \div 1\dfrac{1}{9}$ (c) $\left(\dfrac{9}{10} \times \dfrac{5}{8}\right) \div \dfrac{3}{8}$

5. Evaluate (a) $\left(3\dfrac{2}{3} \div 5\dfrac{1}{2}\right) \div \left(4\dfrac{1}{2} \div \dfrac{3}{4}\right)$ (b) $\left(\dfrac{3}{5} \div \dfrac{1}{3}\right) \div \left(\dfrac{3}{4} - \dfrac{7}{10}\right)$

 Integration Activity

As a member of a Christmas cooperative scheme, Mr. Barnabas contributed $3\dfrac{1}{4}$ times the amount contributed by Mr. Simon.

(ii) How many kilograms of meat will be given to Mr. Barnabas given that Mr. Simon took $2\dfrac{1}{2}$ kilograms of meat?

(iii) How many kilograms of rice will be given to Mr. Simon given that Mr. Barnabas took $47\dfrac{1}{2}$ kilograms of rice?

6.7 Expressing one Quantity as a Fraction of Another

 Example

Express 25 as a fraction of 80

Solution

25 as a fraction of 80 $= \dfrac{\overset{5}{\cancel{25}}}{\underset{16}{\cancel{80}}} = \dfrac{5}{16}$

6.8 Application of Fractions to Real Life Situations

 Real life Example

Out of the 13464 candidates who sat for an examination, 7854 passed. What fraction of the candidates passed?

Solution

Fraction that passed $= \frac{7854}{13464} = \frac{7}{12}$

 Skill Building Exercise 6:8

In the table below express the number in column A as a fraction of that in column B.

	A	B
(a)	20	25
(b)	35	90
(c)	280	150
(d)	525	400

 Real life Exercise

Mrs. Fru and Mr. Lanjo have 25 shares and 30 shares respectively in a company. Given that the total shares in the company is 350 and that all these shares are shared to the shareholders according to their shares. What fraction of the profit will (a) Mrs. Fru take? (b) Mr. Lanjo takes?

Integration Activity

1. Your school registered 400 candidates for an examination and 16 candidates were absent while 280 candidates passed.
   (a) How many students failed
   (b) Calculate the fraction of students who
      (i) passed   (ii) failed   (iii) were absent
   (c) Express as a fraction the number of students who failed to the number who passed.

2. You need two strips of cable that measure $3\frac{5}{6}$ m each. A friend gives you $7\frac{1}{2}$ m. Will this cable be enough?

3. A tailor reduces a bed sheet which is $213\frac{1}{4}$ m long by $5\frac{3}{8}$ m. How long will be the remaining bed sheet?

4. A floor has an area of 69 square metres. Will a carpet $6\frac{1}{8}$ m by $11\frac{1}{4}$ cover the floor?

 Multiple Choice Exercise 6

1. The fractions which is equivalent to $\frac{5}{6}$ is:

   [A] $\frac{7}{6}$ [B] $\frac{2}{3}$ [C] $\frac{30}{36}$ [D] $\frac{11}{12}$

2. The fractions which is not equivalent to $\frac{3}{4}$ is:

   [A] $\frac{1}{4}$ [B] $\frac{12}{16}$ [C] $\frac{30}{40}$ [D] $\frac{33}{44}$

3. In its lowest terms $\frac{60}{108}$ is:

   [A] $\frac{10}{8}$ [B] $\frac{15}{27}$ [C] $\frac{5}{9}$ [D] $\frac{20}{36}$

4. The largest of the following fractions is:
   [A] $\frac{2}{3}$ [B] $\frac{11}{15}$ [C] $\frac{7}{10}$ [D] $\frac{5}{6}$

5. The fractions which is odd is:
   [A] $\frac{3}{4}$ [B] $\frac{7}{8}$ [C] $\frac{4}{3}$ [D] $\frac{1}{5}$

6. $7\frac{4}{5}$ expressed as an improper fraction is:

   [A] $\frac{35}{5}$ [B] $\frac{27}{5}$ [C] $\frac{20}{5}$ [D] $\frac{39}{5}$

7. As a mixed number $\dfrac{40}{3}$ is;

 [A] $13\frac{2}{3}$ [B] $13\frac{1}{3}$ [C] $13\frac{1}{4}$ [D] $13\frac{3}{4}$

8. The sum of $\dfrac{1}{5}$ and $\dfrac{3}{10}$ is :

 [A] $\frac{4}{5}$ [B] $\frac{1}{2}$ [C] $\frac{4}{10}$ [D] $\frac{4}{15}$

9. In its lowest terms $\dfrac{5}{6} - \dfrac{5}{8}$ is :

 [A] $\frac{5}{12}$ [B] $\frac{10}{48}$ [C] $\frac{35}{24}$ [D] $\frac{5}{24}$

10. On simplification $1 - \left(\dfrac{1}{4} + \dfrac{2}{3} \right)$ gives:

 [A] 0 [B] $\frac{11}{12}$ [C] $\frac{1}{12}$ [D] $\frac{4}{7}$

11. A jar is $\dfrac{4}{5}$ full of water. If Nfor drinks $\dfrac{5}{9}$ of the water the fraction of the water left will be:

 [A] $\frac{11}{45}$ [B] $\frac{4}{9}$ [C] $\frac{1}{5}$ [D] $\frac{16}{45}$

12. A man gave $\dfrac{5}{8}$ of his money to his wife and $\dfrac{1}{4}$ of it to his son. The fraction of the money that remains with him is:

 [A] $\frac{1}{6}$ [B] $\frac{1}{4}$ [C] $\frac{1}{3}$ [D] $\frac{1}{8}$

13. $\dfrac{2}{3}$ and $\dfrac{1}{4}$ of a floor are covered by tiles and carpet respectively. The fraction of the floor that is not covered is:

 [A] $\frac{11}{12}$ [B] $\frac{5}{12}$ [C] $\frac{1}{12}$ [D] $\frac{3}{4}$

14. $\dfrac{7}{9}$ of $6\dfrac{3}{7}$ is equal to:

 [A] 5 [B] 10 [C] 12 [D] 15

15. Three quarters of 12 is:

 [A] 16 [B] 8 [C] 7 [D] 9

16. A student ate $\dfrac{1}{2}$ of the $\dfrac{2}{3}$ of the food he preserved for super. The fraction left is:

 [A] $\frac{1}{3}$ [B] $\frac{2}{3}$ [C] $\frac{1}{2}$ [D] $\frac{3}{5}$

17. The quotient $2\dfrac{1}{5} \div \dfrac{1}{5}$ is equal to:

 [A] 10 [B] 12 [C] 11 [D] 13

18. $\dfrac{4}{5}$ of $\left(\dfrac{1}{2} + \dfrac{3}{4} \right)$ has the value of:

[A] 1   [B] 2   [C] 3   [D] 4

19. The product of $\frac{1}{6}$ and the sum of $\frac{2}{5}$ and $1\frac{1}{3}$ is:

[A] $\frac{13}{45}$ [B] $\frac{14}{45}$ [C] $\frac{15}{45}$ [D] $\frac{17}{45}$

20. A labourer's monthly salary is 4800 FRS. If he saves $\frac{1}{5}$ of this amount, in one

year he saves:
[A] 9,600 FRS [B] 115,200 FRS   [C] 180,000 FRS [D] 115,200 FRS

21. A man spends $\frac{3}{4}$ of his monthly salary on food and $\frac{1}{2}$ of the remainder on rent.

If he has 15000 FRS left, the amount he earn is:
[A] 60,000 FRS [B] 90,000 FRS   [C] 105,000 FRS [D] 120,000 FRS

22. Given that a pole has $\frac{1}{3}$ of its length in mud, $\frac{2}{5}$ of the remainder in water and the
rest 6 m long above the surface of the water, the length of the pole is:
[A] 12 m   [B] 10 m   [C] 15 m   [D] 16 m

23. The fraction which must be subtracted from the sum of $2\frac{1}{6}$ and $2\frac{7}{12}$ to

give $3\frac{1}{4}$ is:

[A] $\frac{1}{3}$ [B] $\frac{1}{2}$ [C] $1\frac{1}{2}$ [D] $1\frac{1}{16}$

24. A school girl spends $\frac{1}{4}$ of her pocket money on books and $\frac{1}{3}$ on a

dress. The fraction of her pocket money remaining will be:
[A] $\frac{5}{12}$ [B] $\frac{7}{12}$ [C] $\frac{5}{6}$ [D] $\frac{1}{6}$

25. A man spent $\frac{3}{8}$ of his salary on rent and $\frac{1}{3}$ of the remainder on

cloths. The fraction of his salary that was left, was:
[A] $\frac{23}{24}$ [B] $\frac{5}{6}$ [C] $\frac{19}{24}$ [D] $\frac{5}{12}$

Topic 7

DECIMAL FRACTIONS

Objectives

At the end of this topic, the learner should be able to:

1. Define a decimal as a fraction with a power of ten as the denominator.
2. State the value of a digit in a given decimal.
3. Compare decimals by the use of place value.
4. Convert decimals to fractions and vice versa.
5. Add, subtract, multiply and divide decimals.
6. Multiply and divide decimals by powers of ten, by moving the decimal marker.
7. Round down and round up numbers.
8. Express numbers to a given number of decimal places.
9. Express decimal numbers (whole numbers and decimals) in standard form.

7.1 Notion of Decimal Fractions

A **decimal fraction** or decimal is a fraction whose denominator is a power of 10.

Examples of decimal fractions are $\frac{1}{10}, \frac{348}{100}, \frac{32}{1000}, \frac{257}{10000}, \frac{75}{10}$, etc. It saves time and space to write decimals or decimal fractions as shown in the following examples

$$\frac{1}{10} = 0.1, \quad \frac{348}{100} = 3.48, \quad \frac{32}{1000} = 0.032, \quad \frac{75}{10} = 7.5, \quad \frac{257}{10000} = 0.0257, \quad \frac{75}{10} = 7.5 \text{ etc}$$

The dot is called a **decimal point** or **decimal marker**. In writing decimals using a decimal marker, first count the number of zeros in the power of 10, then imagining the decimal point to be to the right of the last (unit) digit, count the corresponding number of places to the left. Sometimes we write a zero to the left of the decimal point when a number is less than one. For example, we can write $\frac{1}{4}$ as either .25 or 0.25.

 Example

Rewrite the following decimals using a decimal marker.

(a) $\frac{7}{10}$ (b) $\frac{931}{100}$ (c) $\frac{573}{1000}$ (d) $\frac{647}{10000}$

Solution

(a) $\frac{7}{10} = 0.7$ (power of 10 has 1 zero)

(b) $\frac{931}{100} = 9.31$ (power of 10 has 2 zeros)

(c) $\frac{573}{1000} = 0.573$ (power of 10 has 3 zeros)

(d) $\frac{647}{10000} = 0.0647$ (power of 10 has 4 zeros)

Therefore, to divide by a power of 10, count the zeros in the power of 10 and move the decimal point a corresponding number of decimal places to the left.

7.2 Place Value System and Decimals

Each digit to the right of the decimal point represents a number of tenths, hundredths, thousandths etc. as shown in the illustration below.

Place Value of Digits	Million (1,000,000)	Hundred thousand (100,000)	Ten thousand (10,000)	Thousand (1,000)	Hundred (100)	Tens (10)	Units (1)	DECIMAL POINT	Tenth (0.1)	Hundredth (0.01)	Thousand (0.001)	Ten thousandth (0.0001)	Hundred thousandth (0.00001)	Millionth (0.000001)
Number	3	5	2	8	4	1	3	•	6	4	7	1	5	9
Value of Digits	3,000,000	500,000	20,000	8,000	400	10	3		0.6	0.04	0.007	0.0001	0.00005	0.0000009
	Whole Number Part								Decimal Part					

Example

State the value of 3 in each of the following.
(a) 0.03   (b) 0.483   (c) 51.25431

Solution

(a) three hundredths   (b) three thousandths   (c) thirty thousandths

---

Skill Building Exercise 7:1

1. Rewrite the following decimals using a decimal marker.

   (a) $\frac{675}{1000}$ (b) $\frac{586}{100}$ (c) $\frac{389896}{1,000,000}$ (d) $\frac{76587}{10000}$ (e) $\frac{45350}{100,000}$

2. What is the value of the underlined digit?

   (a) 0.<u>6</u>   (b) 0.084<u>2</u>   (c) 0.3254<u>7</u>   (d) 4.8<u>1</u>3

103

7.3 Inter-conversion of Fractions and Decimals

Converting Fractions to Decimals

 Review Exercise

(1) Find the missing number so that the fractions are equivalent.

(a) $\frac{3}{4} = \frac{?}{100}$ (b) $\frac{35}{40} = \frac{?}{100}$ (c) $\frac{70}{125} = \frac{?}{100}$ (d) $\frac{15}{80} = \frac{?}{100}$

(2) Use the knowledge acquired in section 6.1 to write the decimals on the right hand side in (1) using a decimal marker

1. To change a fraction to a decimal, first write the fraction as an equivalent fraction whose denominator is a power of 10, and then write the decimal using a decimal marker.
2. If the fraction is a mixed number first change the fractional part to a decimal and combine the two.

 Example

1. Convert $\frac{4}{25}$ to a decimal.

   Solution

   $\frac{4}{25} = \frac{4}{5} \times \frac{4}{4} = \frac{16}{140} = 0.16$ | Alternatively we can use long division.

   $$\frac{4}{25} = 25\overline{)4}$$

   $$= 25\overline{)4}^{\,0.16}$$

   $$= 0.16$$

2. Convert $2\frac{1}{25}$ to a decimal.

Solution

$$\frac{1}{25} = 25\overline{)\begin{array}{c} 0.04 \\ 100 \\ \underline{100} \\ -\ - \end{array}}$$

$$2\frac{1}{25} = 2.004$$

Alternatively, change the mixed number into an improper fraction before dividing

$$2\frac{1}{25} = \frac{51}{25}$$

$$= 25\overline{)\begin{array}{c} 2.04 \\ 51 \\ \underline{50} \\ 100 \\ \underline{100} \end{array}}$$

Converting Decimals to Fractions

? **Brainstorming Exercise**

1. How many decimal places are there in 0.135?
2. What is the power of 10 which has the same number of zeros as the number of decimal places in 0.135?
3. Write 0.135 as a fraction whose denominator is the power of 10 in (2) and simplify your result.

To change a decimal to a fraction, count the number of decimal places in the number then ignoring the decimal point, divide the number obtained by the power of 10 having the same number of zeros as the number of decimal places. Simplify the result.

 Example

Convert the following decimals to fractions. (a) 0.125   (b)   0.6

Solution

(a) $0.125 = \frac{125}{1000} = \frac{1}{8}$

(b) $0.6 = \frac{6}{10} = \frac{3}{5}$

Skill Building Exercise 7:2

1. Convert the following fractions to decimals.
    (a) $\frac{3}{4}$ (b) $\frac{1}{2}$ (c) $\frac{2}{5}$ (d) $\frac{5}{8}$ (e) $\frac{3}{2}$ (f) $\frac{5}{4}$ (g) $2\frac{1}{4}$ (h) $5\frac{12}{25}$

2. Convert the following decimals to fractions.
    (a) 0.8 (b) 0.65 (c) 1.23   (d) 3.75   (e) 7.5   (f) 2.32

7.4   Operations with Decimals

Addition and Subtraction of Decimals

To add or subtract decimals, arrange the numbers so that the digits of the same place value are in the same column then add or subtract as with whole numbers.

 Example

1. Compute the following without using a calculator.
    (i)   6.04 + 3.23   (ii)   6.4163 + 7.3187 + 5.4128

Solution

(i)   6.04
     +   3.23
         9.27

(ii)   6.4 1 6 3
             7.3 1 8 7
     +   5.4 1 2 8
         19.1 4 7 8

2.   Evaluate (i)   0.957 − 0.831   (ii) 8.90 − 2.47

Solution

(i)   0.9 5 7
     −   0.8 3 1
         0.1 2 6

(ii)   8.9 0
     −   2.4 7
         6.4 3

Skill Building Exercise 7:3

Without using a calculator evaluate the following.
(1) 98.95 + 45.35   (2) 0.89 + 2.083   (3) 6.75 + 8.68 + 8.76
(4) 8.316 + 2.492 + 3.542   (5) 3.238 + 34.2578
(6) 89.7675 + 86.9847   (7) 5.45 + 6.76 + 4.65
(8) 76.743 + 6.467 + 67.58   (9) 7.3 − 0.78
(10) 6.3419 − 2.4834   (11) 8.7234 − 2.6009   (12) 24.88 − 9.48

Multiplying Decimals

 ## Investigative Activity

1. Use a calculator to evaluate the following.
    (a) 26 × 7   (b) 2.6 × 7   (c) 452 × 23   (d) 4.52 × 2.3
    (e) 54.26 × 10   (f) 3.6233 × 100   (g) 0.253 × 1000
2. What conclusions do you draw?

1. *To multiply decimals, multiply as with whole numbers. The number of
    decimal places in the product is the sum of the number of decimal places in
    the numbers.*
2. *To multiply by a power of 10, count the number of zeros in the power of 10,
    and simply move the decimal point a corresponding number of places to the
    right.*

 ## Example

Evaluate the following without using a calculator.
(i) 0.14× 6   (ii) 136.8 × 47   (iii) 0. 2 3 6 × 0.3   (iv) 0. 6 × 0. 8
(v) 10 × 2.581   (vi) 100 × 2.581   (vii) 1000 × 2.581

Solutions

(i)   14
    × 6
    —————
    84
    ═════

Moving 2 decimal places to the left we have 0.84.

(ii)   1368
    ×   47
    —————
    9576
    5472
    —————
    64296
    ═════

Moving 1 decimal place to the left we have 6429.6.

(iii)   0. 2 3 6   3 decimal places
      ×   0.3   1 decimal place
      —————————
      0 .0 7 0 8   4 decimal places

(iv)   0. 6   1 decimal place
    × 0. 8   1 decimal place
    —————
    0.48   2 decimal places

(v) $10 \times 2.581 = 25.81$ Move 1 decimal place to the right.

(vi) $100 \times 2.581 = 258.1$ Move 2 decimal places to the right.

(vii) $1000 \times 2.581 = 2581$ Move 3 decimal places to the right.

⚒ Skill Building Exercise 7:4

Without using a calculator perform the following products

1.   4.5   2.   7.86   3.   5.643   4. 45.65   5. 6.05×45.36
    × 0.3   × 8.97   × 4.37   × 4.37

6. 0.453×0.436 7. 0.065×4.3 8. 6.74×34.23
9. 7.656×4.32 10. 0.035×34.57 11. 6×0.008
12. 3×0.012 13. 100×63 14. 100×9.34
15. 1000×2.75 16. $10,000 \times 0.0065$

Dividing Decimals

To divide decimals, determine which of the dividend or divisor has the greater number of decimal places and count them. Multiply both the dividend and the divisor by an equivalent power of 10, to get rid of the decimals. Simplify or do short or long division as with whole numbers.

Example

Compute the following without using a calculator.

(a) $0.9 \div 3$ (b) $17.022 \div 6$

Solution

We may use long or short division.

(a)
$$
\begin{array}{r}
0.3 \\
3\overline{)0.9} \\
\underline{0.9} \\
--
\end{array}
$$

(b)
$$
\begin{array}{r}
2.837 \\
6\overline{)17.022} \\
\underline{12} \\
50 \\
\underline{48} \\
22 \\
\underline{18} \\
42 \\
\underline{42} \\
--
\end{array}
$$

Alternatively,

(a) $3\overline{)0.9} = \dfrac{0.9}{3} \times \dfrac{10}{10} = \dfrac{\overset{3}{\cancel{9}}}{\underset{1}{\cancel{3}}} \times \dfrac{1}{10} = 0.3$

(b) $6\overline{)17.022} = \dfrac{17.022}{6} \times \dfrac{1000}{1000} = \dfrac{\overset{2\ 8\ 3\ 7}{\cancel{17022}}}{\underset{1}{\cancel{6}}} \times \dfrac{1}{1000} = 2.837$

 Example

Evaluate   (a) $0.5\overline{)3.5}$ (b) $0.00025\overline{)0.4}$

Solution

(a) $0.5\overline{)3.5} = 5\overline{)35}^{\,7}$ (b) $0.00025\overline{)0.4} = 25\overline{)40000}^{\,1600}$

$= 7$ $= 1600$

Skill Building Exercise 7:5

(1) $5\overline{)3.58}$ (2) $25\overline{)0.475}$ (3) $3\overline{)3.9}$ (4) $2\overline{)4.674}$

(5) $7\overline{)3.514}$ (6) $0.025\overline{)0.004}$ (7) $0.5\overline{)6.5}$ (8) $0.05\overline{)0.564}$

(9) $0.06\overline{)23.4}$ (10) $4.3\overline{)0.44564}$ (11) $6.08 \div 4$

(12) $0.551 \div 2.3$ (13) $14.688 \div 4.32$ (14) $1.575 \div 4.5$

7.5 Types of Decimals

The following classification chart shows the classification of decimals.

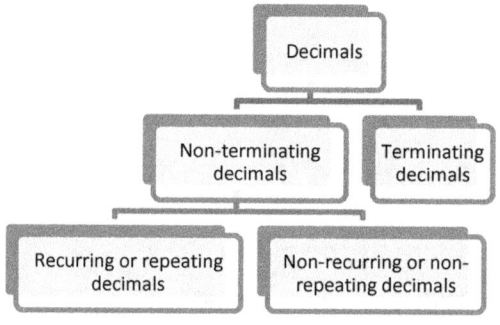

A **terminating decimal** is a decimal, with a finite number of figures. Terminating decimals result from the conversion of fractions such as $\frac{1}{4}, \frac{3}{5}$ and $\frac{7}{20}$ whose denominators when in their lowest term have only 2 or 5 or both as their prime factors. Examples of terminating decimals are 0.25, 1.835 etc. Sometimes on division, the process continues indefinitely. Such decimals are called **non-terminating or recurring or repeating decimals**. For instance

$$1 \div 3 = 3\overline{)1} = 3\overline{)\begin{array}{l} 0.3\overline{3} \\ \hline 10 \end{array}}$$

$$\frac{9}{10}$$

$$\frac{9}{1}$$

$$= 3\overline{3}$$

A **non-terminating decimal** is a decimal with an infinite number of figures. Non-terminating decimals result from the conversion of fractions such as $\frac{4}{7}, \frac{7}{12}$ and $\frac{7}{13}$ whose denominators when in their lowest term have prime factors other than 2 or 5 or both. Examples of non-terminating decimals are $0.333\overline{3}, 3.1415926535898\cdots$, etc.

There are two types of non-terminating decimals. These are recurring or repeating decimals and non-recurring or non-repeating decimals.

A **recurring** or **repeating decimal** is a decimal with an infinite number of figures in which a pattern of one or more digits is repeated indefinitely. Examples of non-terminating decimals are $0.33\overline{3}, 0.7142857\overline{142857}$, etc. The bar placed on any digits show that these digits repeat indefinitely.

A **non-recurring** or **non-repeating decimal** is a decimal with an infinite number of figures in which there is no repetition of a pattern of digits. An example of a non-recurring or non-repeating decimal is $\pi = 3.1415926535898\cdots$, etc.

A **mixed recurring decimal** a decimal with an infinite number of figures in which at least one of the digits after the decimal point does not repeat but some others repeat.

 Skill Building Exercise 7:6

1. Which of the following fractions are likely to result in terminating decimals?
   (a) $\frac{1}{4}$ (b) $\frac{7}{9}$ (c) $\frac{3}{8}$ (d) $\frac{3}{2}$ (e) $\frac{3}{11}$ (f) $\frac{11}{25}$
   (g) $\frac{1}{6}$ (h) $1\frac{3}{40}$ (i) $\frac{4}{7}$ (j) $\frac{3}{20}$ (k) $2\frac{5}{6}$ (l) $\frac{7}{10}$
2. Express each of the fractions in (1) as a decimal.

 Integration Activity

2. The length of a second hand trouser is 103.43 cm. You are a tailor. How many cm would you reduce from this length so that it fits your customer whose length is 98.68 cm?

3. Unity Bank and Booster bank advertise that the interests paid per thousand on money invested are $1\frac{5}{6}$ % and 1.85% per month respectively. In which bank will you prefer to invest your money?

7.6 STANDARD FORM

? Brainstorming Exercise

The mass of an electron is 0.00000000000000000000000000000911 kg.
The Sun is about 150000000 km away from the earth and the closest star to
the Sun is 41600000000000 km away from the sun. This star is called
Proxima Centauri.

3. Read and write each of the numbers in the extract above.
4. How many zeros are there in each of the numbers in the extract? Count
    them.
5. What difficulties have you encountered reading and writing the numbers
    or what are the disadvantages of reading and writing very small or very
    large numbers in this way?
6. Is there no better way in which we can avoid writing so many zeros
    before or after the first digit?
7. What is the first non-zero digit in each of the numbers?
8. In a whole number where is the decimal point (marker) supposed to be?
9. Beginning immediately after the first non-zero digit count the number of
    places up to the decimal point or where the decimal point is supposed to
    be in
    (b) 0.00000000000000000000000000000911
    (c) 150000000
    (d) 41600000000000
10. In which number did you count by moving to the right?
11. In which number did you count by moving to the left?
12. Which of the numbers is greater than 1?
13. Which of the numbers is less than 1?

In science, very large or very small numbers occur as in the extract above.
Numbers written in the form above are said to be in **normal or decimal form**.
Reading and writing numbers in the normal form, especially when they contain
so many zeros is very difficult, consumes time, space and ink and often lead to
so many errors. The **standard form** or **scientific notation** is an easier way of
writing such large or small numbers. The scientific notation is a floating-point
system in which numbers are expressed as products consisting of a number
between 1 and 10 multiplied by an appropriate power of 10. The power of ten is
negative when the number is less than 1 and is positive when the number is
greater than 1.

To express a number in standard form or scientific notation,

(i) Write the number as a product of a number with the decimal point immediately after the first non-zero digit of the given number and a power of ten.

(ii) The power of 10 is obtained by counting the number of decimal places from immediately after the first non-zero digit in the number up to the decimal point or where the decimal point is supposed to be in the case of whole numbers.

(iii) The power of 10 is negative ($-$) if the number is less than 1 and positive ($+$) if the number is greater than 1.

Example

Express the following in standard form
(a) 0.000,000,000,000,000,000,000,000,000,911
(b) 300,000,000   (c) 0,00048   (d) 41600000000000

Solution

(a) $0.000,000,000,000,000,000,000,000,000,911 = 9.11 \times 10^{-28}$
(b) $300,000,000 = 3 \times 10^{8}$ (c) $0,00048 = 4.8 \times 10^{-4}$
(d) $41600000000000 = 4.16 \times 10^{13}$

Skill Building Exercise 7:7

Express the following numbers in standard form.
(a) 5000   (b) 480   (c) 10200   (d) 700000
(e) 0.0032   (f) 0.000073   (g) 0.925   (h) 0.00011
(i) 0.5600   (j) 3000×10^{-8} (k) 19.6×10^{-4} (l) 0.034×10^{-2}

Multiple Choice Exercise 7

1   The reciprocal of 0.02 is:
    [A] 500   [B] 50   [C] 0.5   [D] 0.05
2   The reciprocal of 0.0002 is:
    [A] 50   [B] 500   [C] 5000   [D] 50,000
3   To one decimal place, the reciprocal of 0.625 is:
    [A] 1.6   [B] 0.6   [C] 6.3   [D] 62.5
4   $\dfrac{5}{16}$ as a decimal is:

[A] 0.4125   [B] 0.2125   [C] 0.3125   [D] 0.2725

5   25 out of 200 pineapples are bad. As a decimal, the fraction that is bad is:
    [A] 0.875   [B] 0.185   [C] 0.225   [D] 0.125

6   0.375 as a fraction is:
    [A] $\frac{1}{8}$ [B] $\frac{3}{8}$ [C] $\frac{5}{16}$ [D] $\frac{11}{16}$

7   0.71 is equivalent to:
    [A] $\frac{8}{25}$ [B] $\frac{7}{10}$ [C] $7\frac{1}{5}$ [D] $\frac{71}{100}$

8   5 – 0.003 equals:
    [A] 0.002   [B] 4.003   [C] 4.007   [D] 4.997

9   The value of 4.7–1.9 + 2.1 is:
    [A] 5.9   [B]   4.9   [C]   1.7   [D] 8.7

10 The value of 0.6 × 0.04 is:
    [A] 0.24   [B] 0.64   [C] 0.024   [D] 2.4

11 0.4 × 1.4 equals:
    [A] 56   [B] 5.6   [C] 0.056   [D] 0.56

12 0.2 × 0.4 equals:
    [A] 0.8   [B] 0.08   [C] 8   [D] 0.6

13 0.93 + 0.08 is equals to:
    [A] 1.1   [B] 1.01   [C] 1.11   [D] 0.101

14 0.1 × 0.2 × 0.3 is equal to:
    [A] 0.06   [B] 0.006   [C] 0.05   [D] 0.005

15 The value of $\frac{2.4}{4}$ is:
    [A] 0.6   [B] 6   [C] 60   [D] 2.1

16 The value of 136 × 47 is 6392. The value of 1.36 × 4.7 is:
    [A] 0.6392   [B] 6.392   [C] 63.92   [D] 639.2

17 The value of 136 × 47 is 6392. The value of $\frac{63.92}{13.6}$ is:
    [A] 47   [B] 0.047   [C] 0.47   [D] 4.7

18 Given that 225 × 35 = 7875 then, 22.5 × 0.35 is: equal to:
    [A] 0.07875   [B] 0.7875   [C] 7.875   [D] 78.75

19 0.0063 correct to 2 decimal places is:
    [A] 0.006 [B] 0.01 [C] 0.06 [D] 0.10

20 0.067476 correct to three decimal places is:
    [A] 0.067   [B] 0.065   [C] 0.0647   [D] 0.0648

21 24866 written to the nearest hundred is:
    [A] 24800   [B] 25000   [C] 24900   [D] 24870

22 After evaluating 2.35×0.48, the answer to 2 decimal places is:
    [A] 11.28   [B] 1.128   [C] 1.13   [D] 1.10

23 The value of 3.769÷0.7 to the nearest tenth is:
    [A] 5.41   [B] 5.0   [C] 10   [D] 5.4

24 To the nearest whole number, the result of $\frac{6.6\times1.8}{5.4}$ is:
    [A] 2.2   [B] 3   [C] 2   [D] 22

25 0.000252÷0.007 to two decimal places is:
    [A] 0.04   [B] 0.03   [C] 0.36   [D] 0.40

26 By evaluating $\frac{7+3.32}{9.91-5.11}$, the answer to one decimal place will be:

[A] 21.5   [B] 2.1   [C] 22.0   [D] 2.2

27   $0.44734 \div 0.01$, evaluated to the nearest hundredth is:
    [A] 44.7   [B] 45   [C] 44.73   [D] 44.00

28   $\frac{6.3 \times 6.0 \times 2}{3.6 \times 1.4}$, when simplified, the answer to the nearest ten is:
    [A] 15   [B] 20   [C] 10   [D] 1.5

29   930,000,000 in standard form is:
    [A] 93.0×10^9 [B] 9.3×10^8 [C] 9.3×10^7 [D] 9.3×10^{-7}

30   5238, expressed in standard form is:
    [A] 5.238×10^3 [B] 5.238×10^2 [C] 5.238×10^1 [D] 5.238×10^0

31   Expressed in standard form 435600 is:
    [A] 4.536×10^7 [B] 4.536×10^6 [C] 4.536×10^5 [D] 4.536×10^4

32   When expressed in standard form 2789 equals:
    [A] 2.789×10^{-3} [B] 2.789×10^2 [C] 2.789×10 [D] 2.789×10^3

33   $(0.12)^2$ is equal to:
    [A] 1.44   [B] 0.144   [C] 0.0144   [D] 0.24

34   The value of $\frac{1}{0.2} + \frac{1}{0.25}$ is:
    [A]   9   [B] 4.5   [C] 2.5   [D]   45

35   $78.75 \div 0.35$ is:
    [A] 0.225   [B] 0.25   [C] 22.5   [D] 225

36   The square root of 0.0036 is:
    [A] 0.6   [B] 0.006   [C] 0.06   [D] 0.0006

37   In the number 460.32 the actual value represented by the digit 3 is:
    [A] 3   [B] $\frac{3}{10}$ [C] 30   [D] $\frac{3}{100}$

38   In the number 8.6792 the value of the digit 7 is:
    [A] 70   [B] $\frac{7}{10}$ [C] $\frac{7}{100}$ [D] 700

39   The digit 8 occupies the second place to the right of the decimal point. Its value is:
    [A] 8 hundred   [B]   8 tens   [C] 8 tenth   [D] 8 hundredth

40   The value of $\frac{2000 \times 3000}{40}$ is:
    [A] 150   [B] 150,000   [C] 15,000   [D] 1500

41   In standard form, 52006 can be written as:
    [A] 5.2006×10^3 [B] 5.2006×10^{-4} [C] 5.2006×10^4 [D] 5.2006×10^{-3}

42   120,000 written in standard form is:
    [A] 1.2×10^2 [B] 1.2×10^3 [C] 1.2×10^4 [D] 1.2×10^5

43   The number 36700 written in standard form is:
    [A] 3.67×10^3 [B] 3.67×10^5 [C] 3.67×10^4 [D] 3.67×10^2

44   325,000 in standard form is:
    [A] 3.25×10^6 [B] 3.25×10^5 [C] 3.25×10^{-6} [D] 3.25×10^{-5}

45   0.00562 in standard form is:
    [A] 5.62×10^{-3} [B] 0.562×10^{-2} [C] 5.62×10^{-2} [D] 5.62×10^2

46   Express in standard form 0.0462 is:
    [A] 0.462×10^{-1} [B] 0.462×10^{-2} [C] 4.62×10^{-1} [D] 4.62×10^{-2}

47 0.000834 in standard form is:

[A] 8.34×10^{-4} [B] 8.34×10^{-5} [C] 8.34×10^{3} [D] 8.34×10^{4}

48 0.0000027 in standard form is:

[A] 2.7×10^{6} [B] 2.7×10^{-6} [C] 2.7×10^{5} [D] 2.7×10^{-5}

49 0.000,000,070,2 in standard form is:

[A] -7.02×10^{7} [B] 7.02×10^{-7} [C] -7.02×10^{8} [D] 7.02×10^{-8}

50 Expressed in standard form 0.000,082,3 becomes:

[A] 0.823×10^{5} [B] 0.823×10^{-5} [C] 8.23×10^{-5} [D] 823×10^{5}

51 Written in standard form 0.000370 is:

[A] 3.7×10^{-1} [B] 3.7×10^{-2} [C] 3.7×10^{-3} [D] 3.7×10^{-4}

52 46×900 expressed in standard form is:

[A] 4.14×10^{3} [B] 4.14×10^{5} [C] 4.14×10^{4} [D] 4.14×10^{6}

53 When 4 hours is converted to seconds and expressed in standard form, the result is:

[A] 1.44×10^{4} [B] 1.44×10^{-3} [C] 1.44×10^{3} [D] 1.44×10^{-4}

54 258 km when expressed to mm and expressed in standard form becomes:

[A] 2.58×10^{8} [B] 2.58×10^{7} [C] 2.58×10^{6} [D] 2.58×10^{5}

55 $\frac{8.75}{0.025}$, expressed in standard form is:

[A] 3.5×10^{2} [B] 3.5×10^{-2} [C] 3.5×101 [D] 3.5×10^{-3}

56 Given that $0.000208 = 2.08\times10^{\square}$. the missing number is:

[A] 4 [B] −4 [C] 5 [D] −5

Topic 8

ARITHMETIC PROCESSES

Objectives

At the end of this topic, the learner should be able to:

1. Define a percentage as a fraction with denominator 100.
2. Interconvert percentages, vulgar fractions and decimals.
3. Find a percentage of a given quantity.
4. Express one quantity as a percentage of another.
5. Interpret a ratio as a measure of how large one quantity is compared to another.
6. Express two quantities in a given ratio.
7. Simplify ratios.
8. Determine whether or not two ratios are equal.
9. Define a proportion as a statement of equality between two ratios and refers to two quantities that are varying.
10. Represent and interpret proportional parts.
11. Translate and solve practical problems involving ratios and proportions. e.g. Use ratios to determine best buy.

PERCENTAGES

8.1   Concept of a Percentage

Percent means "per hundred". Thus a percentage is a fraction whose denominator is 100. Examples of percentages are $\frac{25}{100}, \frac{130}{100}$, read '25 percent' and

'130 percent' respectively. It saves time and space to write percentages $\frac{25}{100}$ and $\frac{130}{100}$, as 25% and 130 % respectively.

8.2   Conversion of Percentages, Fractions and Decimals

Changing Percentages to Fractions

Since percentage is a fraction whose denominator is 100, to convert a percentage to a fraction, simply write the percentage as a fraction with denominator 100 and simplify.

 Example

Convert the following percentages to fractions.
(i)   25%   (ii)   66%   (iii)   250%

Solution
(i) $25\% = \frac{25}{100} = \frac{1}{4}$ (ii) $66\% = \frac{66}{100} = \frac{33}{50}$ (iii) $250\% = \frac{250}{100} = \frac{5}{2} = 2\frac{1}{2}$

Changing Fractions to Percentages

To change fractions to percentages, change the given fraction to an equivalent fraction with denominator 100 and then write down the numerator followed by the symbol % or alternatively multiply the given fraction by 100%.

 Example

Change $\frac{3}{4}$ to a percentage.

Solution

$$\frac{3}{4} = \frac{3 \times 25}{4 \times 25}$$

$$= \frac{75}{100} = 75\ \%$$

Alternatively, multiplying the fraction by 100%

$$\frac{3}{4} = \frac{3}{4} \times 100\% = 75\%$$

Exercise 8:1

1. Convert the following percentages to fractions
   (a) 25%   (b) 72%   (c) 82%   (d) 95%   (e) $13\frac{1}{2}$%   (f) $\frac{1}{20}$%
   (g) $7\frac{1}{4}$%   (h) $34\frac{3}{4}$%   (i) 400%   (j) 250%   (k) 600%   (l) 750%

2. Convert the following fractions to percentages
   (a) $\frac{3}{4}$ (b) $\frac{4}{5}$ (c) $\frac{300}{100}$ (d) $\frac{17}{50}$ (e) $\frac{450}{100}$ (f) $\frac{13}{25}$

Changing Percentages to Decimals

 Brainstorming Exercise

1. Explain how you will use the rule for division by powers of 10 to change percentages to decimals.
2. How can you use your ideas in (1) to change decimals to percentages?

Using the rule for division by powers of 10 percentages can be changed to decimals by simply moving the decimal point two places to the left.

✎ **Example**

Convert the following percentages to decimals. (i) 25% (ii) 66% (iii) 250%

Solution
By moving the decimal point two places to the left we have;
(i) 25% = 0.25   (ii) 66% = 0.66   (iii)   250% =2.5

Changing Decimals to Percentages

Since this is the reverse of changing percentages to decimals, simply move the decimal point two places to the right.

Example

Convert the following decimals to percentages. (a) 0.41   (b) 3.42

Solution
By moving the decimal point two places to the right we have;
(a) 0.41=41%   (b) 3.42 = 342 %

 Exercise 8:2

1. Convert the following decimals to percentages
    (a)   0.4   (b)   0.75   (c)   2.5   (d)   40.35
2. Convert the following percentages to decimals
    (a) 20%   (b) 35%   (c)   115%   (d)   250%

8.3   Finding a Percentage of a Given Quantity

To find a percentage of a given quantity, multiply the given quantity by the equivalent fraction whose denominator is 100.

 Example

Evaluate 32% of 200

Solution
$32\% \text{ of } 200 = \frac{32}{100} \times 200 = 64$

 Real life Example

Out of the 312 students who took part in the elections for the school senior prefect, Ngeh had 75%. How many votes did he have?

Solution
$75\% \text{ of } 312 = \frac{75}{100} \times 312 = 234$

Competency Based Exercise 8:1

1. The electoral law in your country requires that at least 31 % of the list for municipal councilors must be women. As the leader of your political party in your constituency, you want to ensure that your list satisfies this condition. The number of councilors required for your council is 25. What is the least number of women that must be in your list?
2. New Life and Sons Enterprise is offering a Christmas discount of 20% for any item bought. How much do you need to buy a pair of shoes marked 12, 000 Frs?

8.4 Expressing One Quantity as a Percentage of Another

To express one quantity as a percentage of another, express the quantity as a fraction of the other and multiply by 100.

 ## Example

How many percent of 500 Francs is 175 Francs?

Solution

$$\frac{175}{500} \times 100 = 35\%$$

 ## Real life Example

A shoe company produced 270 shoes on Monday and 210 on Tuesday. What percentage of the shoes was produced on Monday?

Solution
Total number of shoes produced $= 270 + 210 = 480$

Percentage produced on Monday $= \frac{270}{480} \times 100 = \frac{225}{4} = 56.25\%$

Exercise 8:3

1. Evaluate the following.

    (a) 25% of 60 (b) $33\frac{1}{2}$% of 16500 (c) 0.07% of 8000 (d) 120% of 67500

2. In a conference 60% of the participants are women. Given that there are 90 participants how many men attended the conference?

3. What percent of 25 is 4?

4. Express 0.5 as a percentage of 20.

5. 0.4 is how many percent of 5?

6. Out of the 60 students in a class, 40 passed in a mathematics test. How many percent of the students failed?

RATIOS

8.5   The Concept of Ratios

A ratio is a comparison of two or more quantities which are similar or have the same units.

 Example

In a certain class, there are 30 boys and 40 girls. Write down in four different ways the ratio of number of boys to number of girls in the class.

Solution
Ratio of boys to girls = 30:40 = 3: 4.
In four different ways, ratio of boys to girls = 3: 4 or $\frac{3}{4}$ or 75 % or 0.75.

Note that:

1. To find the ratio of one quantity to another, the two quantities must have the same units.

2. Ratios have no units as the units in the numerator or dividends cancel those in the denominator or divisor.

8.6 Simplification of Ratios

We can simplify a ratio which contains common factors by dividing by the common factors until there are no common factors. Two ratios are equal if they can be written as equivalent fractions.

For instance, $6:8 = 3:4$ since $\frac{6}{8} = \frac{3}{4}$.

 Example

The height of two poles is in the ratio $8:5$. If the second pole is 120 cm what is the height of the first?

Solution
$$\frac{\text{Height of first}}{120} = \frac{8}{5} \Leftrightarrow \text{Height of first} = \frac{8}{5} \times 120 = 192 \text{ cm.}$$

 Competency Based Exercise 8:2

A concrete slab mixture requires 2 bags of cement for 4 wheelbarrows of sand and 3 wheelbarrows of gravel. A builder says he needs 12 wheelbarrows of gravel.
(a) What is the expected number of bags of cement?
(b) What is the expected number of wheelbarrows of sand ?

PROPORTIONS

8.7 The Concept of Proportions

A proportion is a statement involving equal ratios.

Thus $6:8 = 3:4$ or $\frac{6}{8} = \frac{3}{4}$.

The middle terms of a proportion are called the **means** while the outside terms are called the **extremes**. Thus in the proportion $6:8 = 3:4$, 6 and 4 are the extremes while 8 and 3 are the means.

The following formula called the proportion rule is very useful in solving problems involving proportion.

Product of means = Product of extremes

Thus, $6:8 = 3:4 \Leftrightarrow 6(4) = 24$ and $8(3) = 24$

8.8 Proportional parts

$$A \qquad\qquad B \qquad\qquad\qquad\qquad\qquad C$$

Above is a line AC of length 60 mm, which is divided in a ratio 1:3. This means that if the line AB is 1 unit, the line BC is 3 units. If that be the case obviously the line AC must be $1 + 3 = 4$ units. Therefore, the total number of parts is 4. Thus, 60 mm represent 4 parts. The total number of parts is called the **sum of ratio**.

 Example

A woman's salary is 80,000 FRS. The ratio of the money she saved to that spent is 1:3. Calculate the amount (a) saved   (b) spent.

Solution
Sum of ratio $= 1 + 3 = 4$.

(a)   Amount saved $= 80000 \times \frac{1}{4} = 20000$ francs.

(b) Amount spent$= 80000 \times \frac{3}{4} = 60000$ francs.

 Exercise 8:4

1.   During a European Champions League, Eto'o played 6 matches and did not play 4. Express as a ratio:
     (a) The number of matches he played to the total number of matches.
     (b) The number of matches he did not play to the number he played.
2.   A class is made up of 18 boys and 24 girls. Find the ratio of
     (a) Boys to girls
     (b) Boys to the total number of students
     (c) Girls to the total number of students.
3.   Simplify the following ratios (a) 6:9   (b) 5:10   (c) 20:30.
4.   Divide 2000 FRS to two children in the ratio 2:3.
5.   Bih and Mankaa share 70 kg of rice in the ratio of 2:5. What quantity does each get?
6.   Mr. Ngala shares 24000 FCFA to his three sons in the ratio $5:4:3$. How much does each get?
7.   Ndi, Shey and Nfor share 54 oranges in the ratio $2:3:4$. How many does each get?
8.   Given that a man divided his wealth to his three children in the ratio of their

ages 10 years, 15 years and 20 years and that the youngest received 750000 CFA. Find
(a) The total amount
(b) The amount received by each of the elderly children.

Multiple Choice Exercise 8

1. $\frac{3}{40}$ as a percentage is:

   [A] 3%   [B] $7\frac{1}{2}$%   [C] 40%   [D] 75%

2. Expressed as a percentage $\frac{17}{20}$ is:

   [A] 85%   [B] 65%   [C] 55%   [D] 45%

3. $\frac{9}{10}$ as a percentage is:

   [A] 0.9%   [B] 9%   [C] 99%   [D] 90%

4. $37\frac{1}{2}$% expressed as a fraction is:

   [A] $\frac{7}{8}$ [B] $\frac{5}{8}$ [C] $\frac{3}{8}$ [D] $\frac{1}{8}$

5. 0.125 as a percentage is:
   [A] 125%   [B] 12.5%   [C] 1.25%   [D] 0.125%

6. 65 % as a decimal is equivalent to:
   [A] 0.065   [B] 65.0   [C] 6.5   [D] 0.65

7. As a decimal l72% to 2 decimal places is:
   [A] 1.68   [B] 1.70   [C] 1.72   [D] 1.66

8. Ambe's monthly salary is 300.000 FRS given that he spends 15 % on rents. The amount spent on rent is:
   [A] 60,000 FRS   [B] 45,000 FRS   [C] 40,000 FRS   [D] 30,000 FRS

9. A boy scored 70% in a test. If the maximum mark was 40, then the boy's mark was:
   [A] 4   [B] 10   [C] 28   [D] 30

10. If 20 % of a sum of money is 4000 FCFA, then the whole sum of money is:
    [A] 200,000 FCFA [B] 80,000 FCFA [C] 20,000 FCFA [D] 40,000 FCFA

11. In a certain class, 21 students are boys and 19 are girls. The percentage of the class who are boys is:
    [A] 21%   [B] 40%   [C] 47%   [D] 52.5%

12. The percentage of 4 that is 5 is:
    [A] 20%   [B] 80%   [C] 120%   [D] 125%

13. Given that out of the 500 students who sat for an examination 150 students failed. The percentage that failed is:
    [A] 70%   [B] 30%   [C] 60%   [D] 40%

14. A girl obtains 60 marks out of 75 on an examination. This is equivalent to a percentage of:
    [A] 60%   [B] 30%   [C] 90%   [D] 80%

15. The following table shows the means by which children of a certain class travel to school. The percentage that walk to school is:

Means of transport	Percentage
Bike	36%
Bus	20%
Car	19%
Walk	?

[A] 90%   [B] 20%   [C] 25%   [D] 10%

16. Given that $\frac{5}{8}$ of the pupils in a school are boys, then the ratio of boys to girls is:
    [A] 5:3   [B]   5:8   [C] 3:5   [D] 8:5

17. 63000 FCFA is divided into two parts so that the first part is $\frac{3}{4}$ of the second. The value of the larger part in FCFA is:
    [A] 24,000   [B] 27,000   [C] 28,000   [D] 36,000

18. The ratio of A's share to B's share in the profit of a business is 5:4. If the total profit is 450, 000 FCFA, then A's share in FCFA is:
    [A] 90,000   [B] 200,000   [C] 225,000   [D] 250,000

19. If 120,000 FCFA is divided in the ratio 2 to 3 then the smaller share in FCFA is:
    [A] 48,000   [B] 80,000   [C] 72,000   [D] 60,000

20. The ratio of the shares of two partners *A* and *B* in the profit of a business is 5: 3. When the profit is 120,000 FRS *B* will receive:
    [A] 30,000 FRS   [B] 36,000 FRS   [C] 45,000 FRS   [D] 72,000 FRS

21. The ratio of the shares of two partners A and B in the profit of a business is 5:3. When the profit is 48,000 FRS *B* will receive:
    [A] 80,000 FRS   [B] 30,000 FRS   [C] 28,800 FRS   [D] 18,000 FRS

22. A sum of money was divided in the ratio of 2:3 between two children. The fraction of the money received by the child who received the smaller amount is:
    [A] $\frac{2}{5}$ [B] $\frac{3}{5}$ [C] $\frac{1}{2}$ [D] $\frac{2}{3}$

23. The ratio of the number of men to the number of women in a 20-member committee is 3:1. The number of women who must be added to the 20-member committee so as to make ratio of men to women 3:2 is:
    [A] 2   [B] 5   [C] 7   [D] 9

24. If three children share 10500 FRS among themselves in the ratio 6:7:8. The largest share is:
    [A] 3,000 FRS   [B] 3,500 FRS   [C] 4,000 FRS   [D] 4,500 FRS

25. An amount of money is shared among Patience, Quiniva and Rita such that for every 3 Patience gets, Quiniva gets 2 and for every 1 Quiniva gets Rita gets 4. The ratio of Patience to Quiniva to Rita is:
    [A] 3:1:4   [B] 3:2:1   [C] 3:2:4   [D] 3:2:8

26. If the ratio of cement to gravel in a mixture is 4: 9 and the ratio of gravel to

sand is 3: 5, then the ratio of cement to gravel to sand in the mixture is:
[A] 4:9:15   [B] 3:9:14   [C] 4:9:20   [D] 4:12:45

27. The sum of 2400 FCFA is shared between Neba, Ndeh and Ambe. If Neba receives 600 FCFA and the remaining amount is shared between Ndeh and Ambe in the ratio $4 : 5$ respectively. In FCFA Ambe receives:
[A] 800   [B] 900   [C] 1,000   [D] 1,200

Module 2

Introduction to Plane Geometry

Family of Situations

At the end of module 2, the student is expected to be competent within the families of situations 'Representation and transformation of Plane Shapes within the Environment'.

Categories of Action

The categories of action for module 2 include:
1. Perception of the physical environment,
2. production of plane shapes,
3. transformation of the physical environment
4. Determination of measures.

Credit

The module is expected to be covered within 12 weeks teaching 4 hours per week (or within 45 to 48 hours).

Topic 9

POINTS AND LINES

Objectives

At the end of this topic, the learner should be able to:

1. State some uses of points and lines in everyday life.
2. Define a point, a line and a plane.
3. Distinguish between straight lines and curve lines.
4. Distinguish, draw, represent and denote lines, line segments and rays.
5. Identify collinear points and coplanar points.
6. Measure line segments of given lengths.
7. Determine the most suitable unit for measuring a given length or distance.
8. Locate the midpoint of a given line segment.
9. Construct the bisector of a given line segment.
10. Identify parallel lines and perpendicular lines.
11. Construct parallel lines and perpendicular lines.

9.1   Meaning of Geometry

The term *geometry* comes from two Greek words *geô*, which means "earth," and *metrein*, which means "to measure". Thus, geometry is a branch of mathematics that deals with shapes and their sizes in space.

9.2   Brief History of Geometry

The history of geometry can be traced as far back as 1900 BC. The first important geometer is Thales of Miletus, a Greek who lived about 600 BC. Pythagoras and his associates proved many new Geometrical theorems. About 300 BC Euclid a Greek Mathematician wrote a book '*Elements',* on the principles of geometry and properties of numbers. Archimedes, one of the greatest Greek scientists, equally made a number of important contributions to geometry during the 3rd century BC.

9.3   Uses of Geometry

1.   Geometry is very important in everyday life as it helps us to determine properties such as the areas and perimeters of two-dimensional shapes and the surface areas and volumes of three-dimensional shapes. In order to estimate the amount of carpet floor will take or the amount of paint required to paint a house or even the amount of water a tank can contain basic geometrical knowledge is required.
2.   Geometry is also used in survey, navigation and astronomy.
3.   Today in space travel and aviation, geometry is highly needed.

9.4   Basic Geometrical Concepts

The terms **point** and **line** are the most basic geometrical concepts. These terms are further used to develop concepts such as surface, planes and solid.

A **point** usually designated by a capital letter such as *P* next to a dot, has no size and represents a position. For instance on maps a dot is used to represent say a town but it is not the town.

• *P*

A **line** is the path of a series of points. It has a length but no thickness, and may be straight (as in (a), (b) and (c)), curved (as in (d) and (e)) or broken (as in (c) and (e) below). A line is often designated by two capital letters representing any two of its points (as in (a), (c) and (d)) or by a small letter (as in (b) and (e) below).

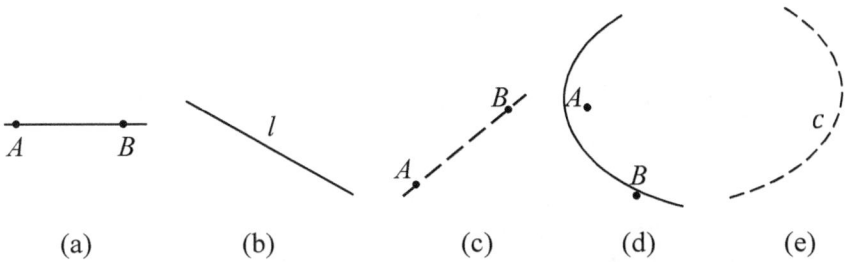

| (a) | (b) | (c) | (d) | (e) |

A line containing two points A and B extends indefinitely and is denoted by (AB).

A **straight line** is the shortest distance between two points. A straight line should be looked at as the path of a point moving in the same direction.

A **curved line** on the other hand is the path of a moving point whose direction is continuously changing.

Line Segment

A **line segment** (figure (a) below) is part of a line consisting of two points, called endpoints, and all the points that lie between these endpoints.

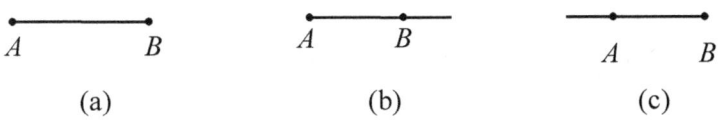

| (a) | (b) | (c) |

A line segment with endpoints A and B is denoted by $[AB]$ or $\overline{AB}$. The length of a line segment $[AB]$ is denoted as AB. Thus if the length of the line segment $[AB]$ is 4 cm, then we can write $AB = 4$ cm.

Half Lines or Ray

A **half line** or a **ray** is a line which has a definite beginning but no definite end. Rays are often designated according to their beginning points and a point through which they pass; a ray with beginning point A that passes through a point B (Figure (b) above) is denoted by $[AB)$ or $\overrightarrow{AB}$ while a ray with beginning point B that passes through point A (Figure (c) above) is denoted by $(AB]$ or $\overleftarrow{AB}$.

Planes

A plane is a surface such that any straight line connecting any two of its points lies entirely in it. A plane has length and width but no depth. A plane is similar to an infinitely large tabletop that has no edges. There exists one and only one plane that contains any three points that do not lie on a single straight line.

131

Group Work

1. Write down four objects bounded by straight lines.
2. Write down four objects bounded by curved lines.
3. Which of the following objects are bounded by
   (a) Straight lines?   (b) Curved lines?
   a ruler, a textbook, a ball, a protractor, the walls of your classroom, an unused piece of chalk, the moon, a star, the sun, a wire, a wheel.

 Exercise 9:1

1. State the meaning of the each of the following notations.
   (a) AB (b) $(AB]$ (c) $[AB)$ (d) (AB) (e) $[AB]$ (f) $\overline{AB}$
2. Using the number line below, find by measurement
   (a) AB (b) BD (c) AC (d) AD (e) $AB + CD$

3. Identify and name each of the following using the appropriate notation.

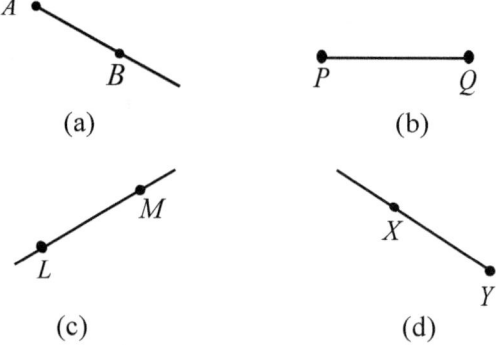

   (a)   (b)

   (c)   (d)

4. How many line segments bound each of the following figures?

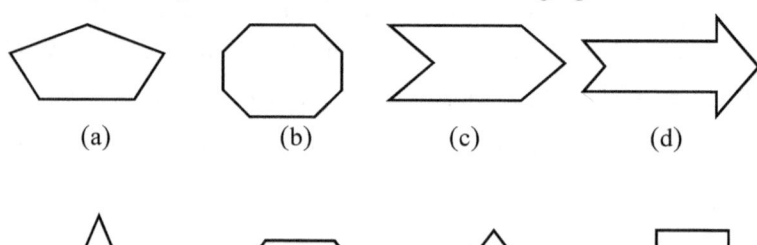

   (a)   (b)   (c)   (d)

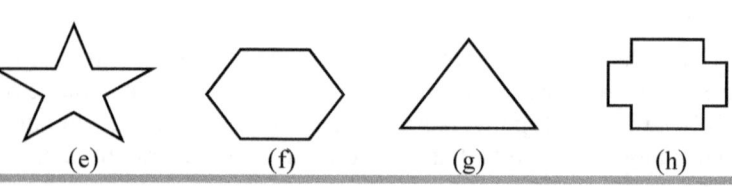

   (e)   (f)   (g)   (h)

9.5   Collinear and Non-Collinear Points

If three or more points lie on a single straight line they are said to be **collinear**. Otherwise they are **non-collinear**. In the figures below, *X*, *Y* and *Z* are collinear while *P*, *Q* and *R*, are non-collinear

 Exercise 9:2

1. How many straight lines are there in figure (i) below? List all the sets of collinear points on each straight line.

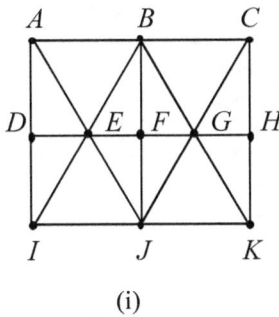 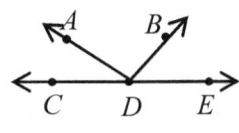

| (i) | (ii) |

2. In Figure (ii) above name:
   (a) Three collinear points.   (b) Three non-collinear points.
   (c) Two rays.   (d) Five line segments.

9.6   Measuring Line Segments

Figure (a) and (b) below, shows how we can measure the line segment *AB* using a pair of dividers and a ruler. Place one of the pins of the pair of dividers at *A* and open it so that the other is at *B*. Carefully take it to your ruler so that one pin is at the zero mark. Read the length where the other pin falls.

Make sure you don't get confused between the inches side and the millimetres (mm) and centimeters (cm) side of your ruler.

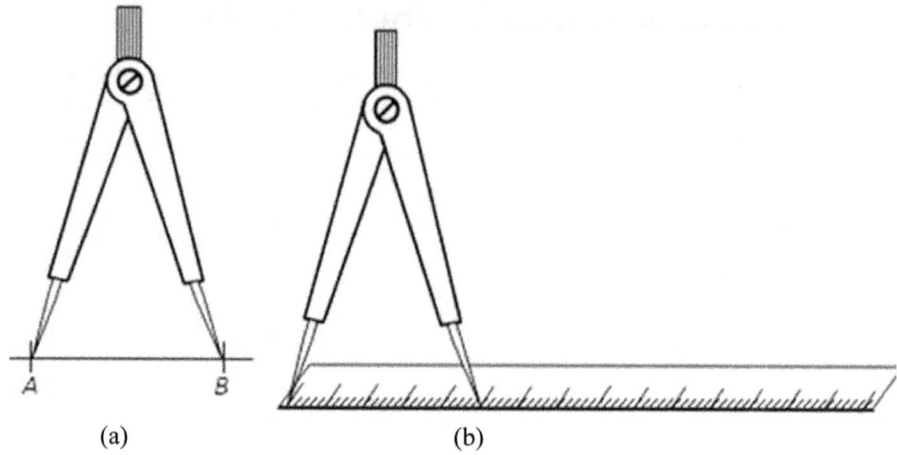

(a)   (b)

 Brainstorming Exercise

1. Is it reasonable to measure distances on say a plot or road using a ruler?
2. Are mm and cm suitable for such distances?
3. What type of instruments could be used in measuring such distances?
4. What units could be used for such distances?
5. Name some other instruments used in measuring length and distance.
6. Is it reasonable to measure lengths on say a paper using a tape?

We can measure short lengths like those on paper using a ruler in mm or cm. Distances like those on a plot or field may be measured in metres (m) using a tape but it would be more reasonable to measure distances on the road in kilometers (km).

Exercise 9:3

1. In the figure below, *ABC* is a line. Measure (a) [*AB*] (b) [*AC*]   (c) [*BC*]

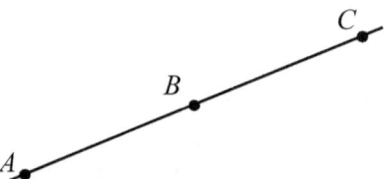

2. In what units will you measure:
    (a) The length of your classroom   (b) The length of your school field.
    (c) The length of the road from your school to the closest town.
    (d) The length of your exercise book.   (e) The length of your desk.

9.7 Perpendicular lines

Perpendicular lines are straight lines which are such that every point on one is equidistant from two points on the other. If two lines L_1 and L_2 are perpendicular we write $L_2 \perp L_2$. To represent two perpendicular lines on a plane, we use a square as shown below.

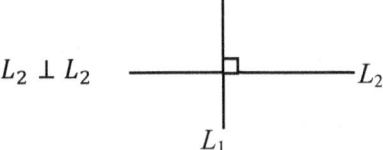

$$L_2 \perp L_2$$

L_2

L_1

 Discussion Exercise

1. Name three things in your classroom which have perpendicular lines and show them.
2. Most buildings have perpendicular corners. Why do you think this is necessary?

The adjacent sides of a desk are perpendicular. The adjacent sides of your books are perpendicular. The vertical lines at the corner of a wall and a horizontal line between the wall and the floor are perpendicular.

9.8 The Midpoint and Mediator of a Line Segment

 Investigative Activity

1. Measure the length of the line segments in the figure below and record your result on the table provided.

Line segment	Length
AM	
MB	
AB	

A M B

0   3   6

2. What conclusion do you draw concerning
    (a) AM and MB? (b) AM and AB? (c) MB and AB? (b) $AM + MB$?

The midpoint of a line segment is the point that divides the line segment into two equal segments. In the figure above, M is the midpoint of the line segment AB.

Generally,

if $M \in [AB]$ such that $AM = MB$ and $AM + MB = AB$, then M is the midpoint of $[AB]$.

Conversely,

if M is the midpoint of $[AB]$, then $AM + MB = AB$.

The **mediator** or **perpendicular bisector** of a given line segment is a line which is perpendicular and passes through the midpoint of the line.

Constructing the Mediator of a Given Line Segment [AB]

? Brainstorming Exercise

1. A student draws two intersecting circles with a pair of compass and a short pencil well fastened to it. If the compass was not altered after drawing the first circle, what can you say about the two circles?
2. How can you use a pair of compass to draw two arcs which intersect at a point which is equal in distance from two given points A and B?

The midpoint of a line segment AB can be determined using a pair of compass and a pencil as follows.

(a) With center A, draw two arcs of equal radii on the two sides of AB.

(b) With center B, draw two arcs of equal radii as those in (a) to cut the first two at C and D.

(c) Now join the points C and D of intersection of these arcs with ruler and pencil. The line CD is the **mediator** or **perpendicular bisector** of the line segment $[AB]$ and passes through the midpoint of $[AB]$.

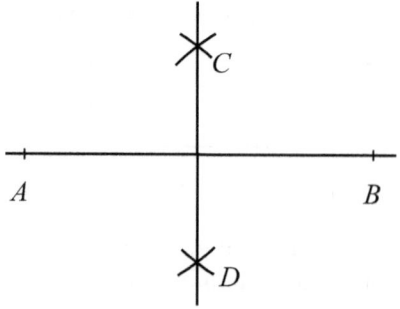

Constructing a **Perpendicular to a Given Line AB, from a Given Point P**

 Example

Construct a perpendicular from the point *P* to the line segment *AB* in figure (a) below.

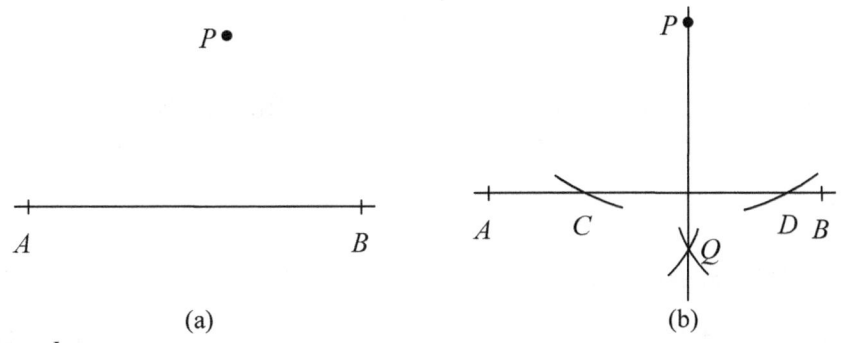

(a)   (b)

Procedure

(i)   With center *P* draw 2 arcs of equal radius to cut *AB* at two points *C* and *D* as in figure (b) above.

(ii) With centers *C* and *D* draw two arcs of equal radius to intersect on the opposite side of *P*. Name this point of intersection *Q*.

(iii) Now join *PQ*. *PQ* is perpendicular to *AB*.

 Discussion Exercise

1.   In building construction, what tools, are normally used in place of a pair of compass to ensure that walls are perpendicular?

2.   Explain how these tools are used.

In building construction, the tools normally used to ensure that walls are perpendicular are the spirit level, a plumb line or a set square.

Exercise 9:3

1. In the figure below, [*XY*] is a line segment. Mark a point *M* where *M* is the midpoint of [*XY*].

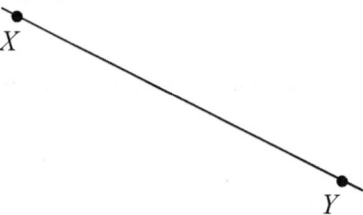

2. Draw a straight-line segment [*CD*] of length 8 cm. Bisect the line [*CD*].

9.9 Parallel Lines

Two lines are said to be **parallel** if they have the same direction. Parallel lines never meet. The following diagram shows that the lines *PQ* and *RS* are parallel.

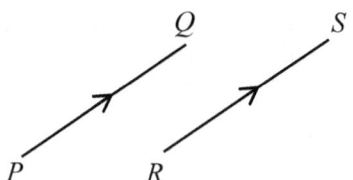

 Discussion Exercise

1. Name three things in your classroom which have parallel lines and show them.
2. The opposite walls of most buildings are parallel. Why do you think this is necessary?

Construction of Parallel Lines

 Brainstorming Exercise

A student draws two circles which do not touch with a pair of compass and a short pencil well fastened to it. After drawing the circles, the student draws a straight line to touch the two circles.

(a) If the compass was not adjusted after drawing the first circle, what can you say about the distance of the line from the center of each circle?

(b) What is the relationship between the line joining the centers of the circles and the line touching the circles?

Example

1. Construct a line parallel to the line *AB* (figure (a) below) and 4 cm away from it.
2. Construct a line parallel to the line *l* (figure (b) below) and passing through the point *P* shown below.

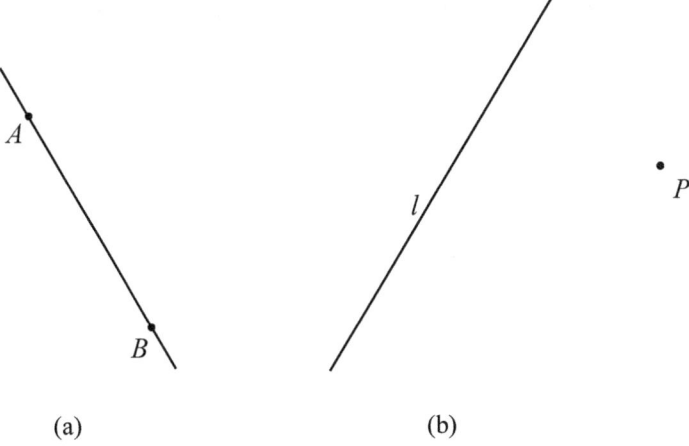

(a)   (b)

1. **Procedure**
    (a) With open pair of compass, measure 4 cm from a ruler.
    (b) With centres *A* and *B* trace the arcs *X* and *Y* respectively on one side of the line *AB*.
    (c) Now draw the line XY to touch the arcs *X* and *Y*.

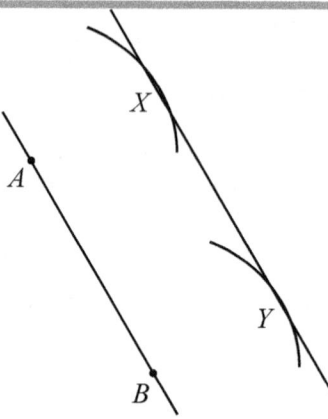

2. **Procedure**
   (a) Choose any point X on l and with XP as radius draw two arcs to pass through P and cut the line l at Y.
   (b) With centre Y draw another arc Z of the same radius and on the same side of l as P. With centre Y draw another arc of radius PY to pass through P. With centre X draw another arc with radius PY to cut Z. Now draw a straight line to pass through P and Z. This is the required line.

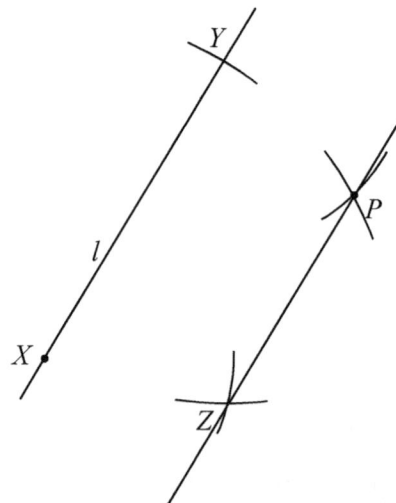

Multiple Choice Exercise 9

1. The figure below shows:

   [A] $\overrightarrow{CD}$ [B] $\overleftrightarrow{CD}$ [C] $\overleftarrow{CD}$ [D] $\overline{CD}$

   C   D

2. The figure below shows:

   [A] $\overrightarrow{CD}$ [B] $\overleftrightarrow{CD}$ [C] $\overleftarrow{CD}$ [D] $\overline{CD}$

   C   D

3. The figure below shows:

   [A] $\overrightarrow{CD}$ [B] $\overleftrightarrow{CD}$ [C] $\overleftarrow{CD}$ [D] $\overline{CD}$

   C   D

4. The figure below shows:

   [A] $\overrightarrow{CD}$ [B] $\overrightarrow{CD}$ [C] $\overleftarrow{CD}$ [D] $\overline{CD}$

   C   D

5. The figure below shows:

   [A] (CD)   [B] [CD)   [C] (CD]   [D] [CD]

   C   D

6. The figure below shows:

   [A] (CD)   [B] [CD)   [C] (CD]   [D] [CD]

   C   D

7. The figure below shows:

   [A] (CD)   [B] [CD)   [C] (CD]   [D] [CD]

   C   D

8. The figure below shows:

   [A] (CD)   [B] [CD)   [C] (CD]   [D] [CD]

   C   D

9. The diagram in the figure below represents:
   [A] A line through the points *A* and *B*
   [B] A ray from the points *A* passing through the point *B*
   [C] A ray from the points *B* passing through the point *A*
   [D] A line segment from the point *A* to *B*

   A   B

10. In the figure (a) below, the points which are collinear with *D* and *R* are:

[A] *F* and *A* [B] *Q* and *A* [C] *Q* and *P* [D] *F* and *P*

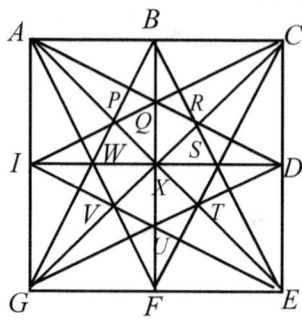

(a)   (b)

11. In figure (b) above, a set of collinear points are:
   [A] *S, Q, R*   [B] *R, Q, P*   [C] *S, P, T*   [D] *T, R, Q*
12. A line is drawn with 2 distinct points *A* and *B*. The maximum number of
    different rays that pass through *A* and *B* are:
   [A] 2   [B] 8   [C] 6   [D] 4*AB* bisects *PQ* at
   point *N*. The statement that is true of *N* is:
   [A] *N* is the midpoint of AB.
   [B] *N* is both the midpoint of *AB* and the midpoint of *PQ*.
   [C] *N* is the midpoint of *PQ*.
   [D] *N* divides *PQ* in the ratio 2:1.
13. Point *P* is the midpoint of AB. Complete the statement: *PB* = 7 cm, *AB* is equal
    to:
   [A] 7 cm   [B] 14 cm   [C] 3.5 cm   [D] none of the above
14. In Figure (a) below, given that *M* is the midpoint of *BC*, *PM* is called the:
   [A] median   [B] mediator   [C] altitude   [D] angle bisector

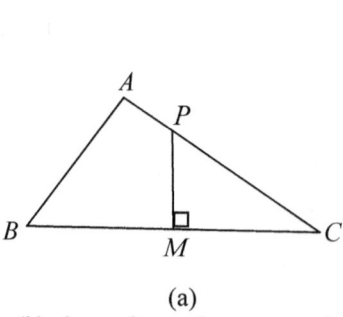

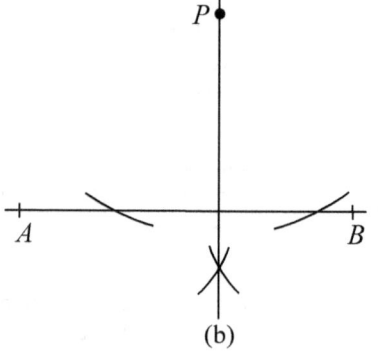

(a)   (b)

15. Figure (b) above shows the construction of:
   [A] A mediator   [B] A congruent angle
   [C] An angle bisector   [D] Intersecting lines
16. Another name for perpendicular bisector is:
   [A] Altitude   [B] Midpoint   [C] Angle bisector   [D] Mediator

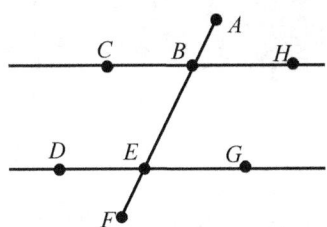

17. In figure above, four line segments with B as endpoint are:

    [A] $\overline{BA}, \overline{BF}, \overline{BC}, \overline{BH},$ [B] $\vec{BA}, \vec{BF}, \vec{BC}, \vec{BH}$

    [C] $\overleftrightarrow{AF}, \overleftrightarrow{CH}, \overleftrightarrow{AE}, \overleftrightarrow{BE}$ [D] $\overline{AF}, \overline{CH}, \overline{AE}, \overline{BE}$

18. In figure above, four rays that have point B as endpoint are:

    [A] $\overleftrightarrow{AB}, \overleftrightarrow{FB}, \overleftrightarrow{CB}, \overleftrightarrow{HB}$ [B] $\vec{BA}, \vec{BF}, \vec{BC}, \vec{BH}$

    [C] $\overline{AB}, \overline{FB}, \overline{CB}, \overline{HB}$ [D] $\overline{BA}, \overline{BF}, \overline{BC}, \overline{BH}$

Topic 10

ANGLES

Objectives

At the end of this topic, the learner should be able to:

1. Understand the meaning of an angle.
2. Understand the meaning of a full turn, a half turn a quarter turn etc.
3. Distinguish between the different types of angles such as an acute angle, a right angle, an obtuse angle, a straight angle and a reflex angle.
4. Name and write angles using letter notation.
5. Identify vertically opposite angles, adjacent angles and angles on a straight line.
6. Use a protractor to measure angles.
7. Bisect a given angle.

10.1 The Notion and Units of an Angles

There are so many objects around us, which turn. Examples are; the wheel of a car, the hands of a clock, the door when it is opened or closed. Think of some more things that turn. Two things of interest to us when things turn are; the point at which the turning takes place called the **center of rotation** and the amount of turn called the **angle of rotation**. The amount of turn or angle of rotation is measured in units called **degrees** and **minutes, radians** and **grades**. For now, we shall concentrate on the degrees and minutes and use it to illustrate the various types of angles.

10.2 Types of Angles

A Full Turn or a Revolution

Consider the movement of the hands of a clock. In one hour, the minute hand turns round once as shown in figure below. This is called a **full turn** or a **revolution.** A full turn is equal to 360 degrees. The symbol for a degree is °. Thus, a **full turn** or a **revolution** is equal to 360°.

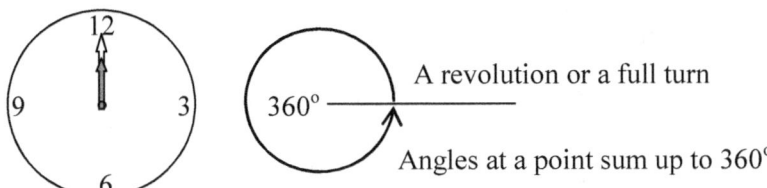

A revolution or a full turn

Angles at a point sum up to 360°

A Half Turn

Using the above idea, through how many degrees will the minute hand turn in 30 minutes (say from 12 to 6 as shown below)?

To answer this question, first determine the fraction of 60 minutes that is 30 minutes. Clearly, the fraction is $\frac{1}{2}$.

Therefore, in 30 minutes, the minute hand will turn half of its journey round. This is called a half turn.

Since a full turn is 360°, a half turn should therefore be $\frac{1}{2}$ of 360° = 180°.

Therefore a half turn is equal to 180° . Another name for a half turn is **angles on a straight line**. The figure below illustrates a straight angle.

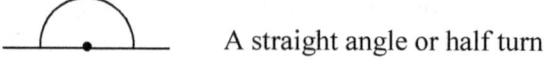

A straight angle or half turn

A Quarter Turn

Again from the above in 15 minutes, the minute hand will turn one quarter of its journey round. This is called a quarter turn. A quarter turn is therefore equal to $\frac{1}{4}$ of 360° = 90°. Another name for a quarter turn or 90° (shown below) is a **right angle**.

On diagrams a right angle is often represented with a small square for the sake of recognition.

Angles which sum up to 90° are called **complementary angles**.

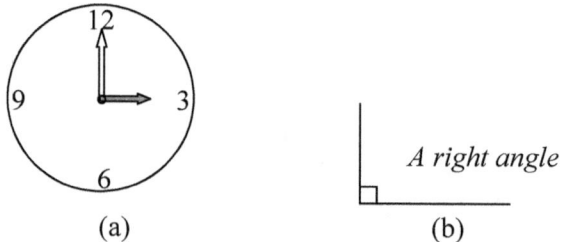

A right angle

(a)   (b)

Two rays lying on the surface of a plane and emanating from the same point form an angle. These rays are the sides of the angle. The following are the various types of angles.

A Zero Angle

A zero angle is an angle which is equal to 0°. This can be looked upon as the angle between two lines drawn so that one is lying on top of the other.

A zero angle

An Acute Angle

An acute angle is an angle which is between 0° and 90°.

An Obtuse Angle

An obtuse angle is an angle, which is greater than 90° and less than 180°.

A Reflex Angle

A reflex angle is an angle, which is greater than 180°.

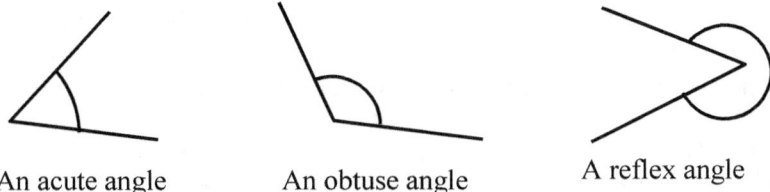

An acute angle   An obtuse angle   A reflex angle

Note: Acute, right, and obtuse angles are the most commonly used angles.

Exercise 10:1

1. State the angle between the following hands of the clock.

(a)   (b)   (c)   (d)

(e)   (f)   (g)   (h)

(i)   (j)   (k)   (l)

2. What fraction of a revolution is:
    (a) 180°   (b) 90°   (c) 60°   (d) 30°
    (e) 45°   (f) 135°   (g) 300°   (h) 240°
3. What fraction of a half turn is (a) 90°   (b) 60°   (c) 30°   (d) 45°
4. What fraction of a half turn is (a) $22\frac{1}{2}°$ (b) 10°   (c) 18°   (d) 42°
5. Determine the angle through which the hour hand of a clock moves between:
    (a) 1:00 p.m. and 7 p.m.   (b) 9:00 a.m. and 12 p.m.
    (c) 2:00 a.m. and 4 a.m.   (d) 11:00 a.m. and 4:30 p.m.
6. Through what angle does the minute hand of a clock move between:
    (a) 2:15 p.m. and 3:15 p.m.   (b) 4:15 p.m. and 5:25 p.m.
    (c) 10:45 a.m. and 10:50 a.m.   (d) 7:15 p.m. and 7:45 p.m.
7. State the angle between the hour and minute hands of a clock at:
    (a) 12 O'clock   (b) 9 O'clock   (c) 6 O'clock   (d) 4 O'clock
8. Classify the following angles as acute, right, obtuse, straight, reflex or revolution.

    (a)   (b)   (c)   (d)   (e)

    (f)   (g)   (h)   (i)   (j)

9. Which of the following angles are acute, right, obtuse, straight, reflex angles or revolutions?
    (a) 140°   (b) 80°   (c) 250°   (d) 180°   (e) 16°   (f) 170°
    (g) 360°   (h) 110°   (i) 54°   (j) 100°   (k) 330°   (l) 150°
10. Give one value for an angle that is:
    (a) acute   (b) right   (c) a revolution   (d) straight   (e) a reflex   (f) obtuse

10.3 Standard Notation for Angles

Consider the angle formed by the rays [*AB*) and [*AC*) both emanating from the point *A*.

The various ways of naming the angle in the diagram below are Angle *A*, angle *a*, ∠*BAC*, ∠*CAB*, angle *BAC*, or angle *CAB*.

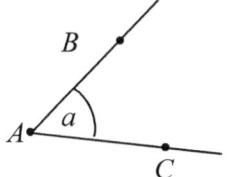

10.4 Measuring Angles

1. What is the name of the instrument below?
2. What is it used for?
3. Describe the instrument and state what the numbers stand for.
4. Discuss the precautions you will need to take when using the instrument.

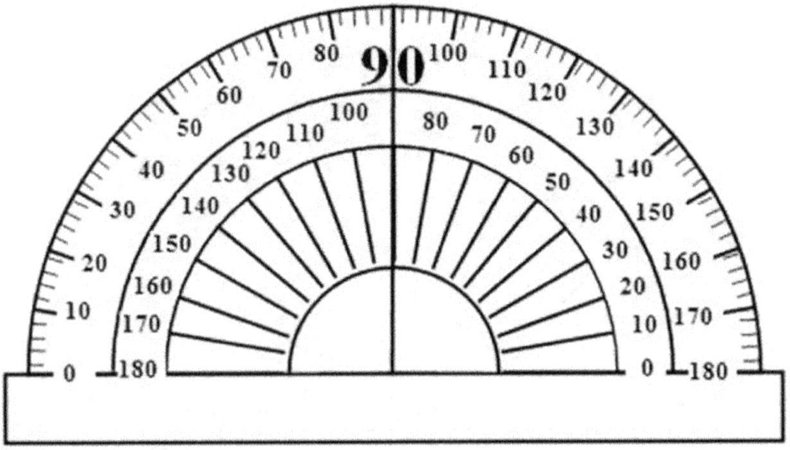

The instrument used for measuring angles is a protractor. Most protractors are graduated from 0° to 180°on two scales, which are the same but arranged oppositely. The center of the protractor is where the symmetrical vertical line meets the straight edge. To measure angles with a protractor,

1. We ensure that the center of the protractor is at the point of intersection of the lines and that one of the lines lie exactly under the straight edge.
2. We read the angle using the scale whose zero value is on the straight edge which lies on top of the line.
3. If the lines which form the sides of the angles are too short, extend them so that the protractor does not cover them completely.

Example

Measure the angle *XOY* in the figure below.

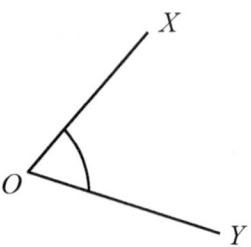

Solution
There are two ways of placing the protractor. Place the protractor so that the straight edge is either along the line *OY* (as in figure (a)) or along the line *OX* (as in figure (b)). In either ways, the center of the protractor is at *O* and the angle is within the protractor. By measurement, the angle is approximately 68°.

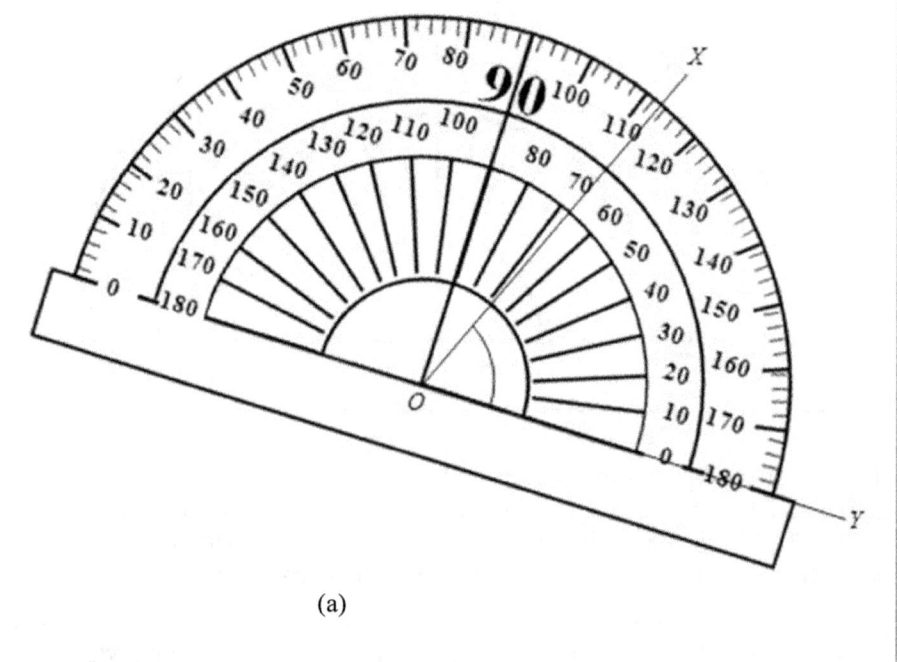

(a)

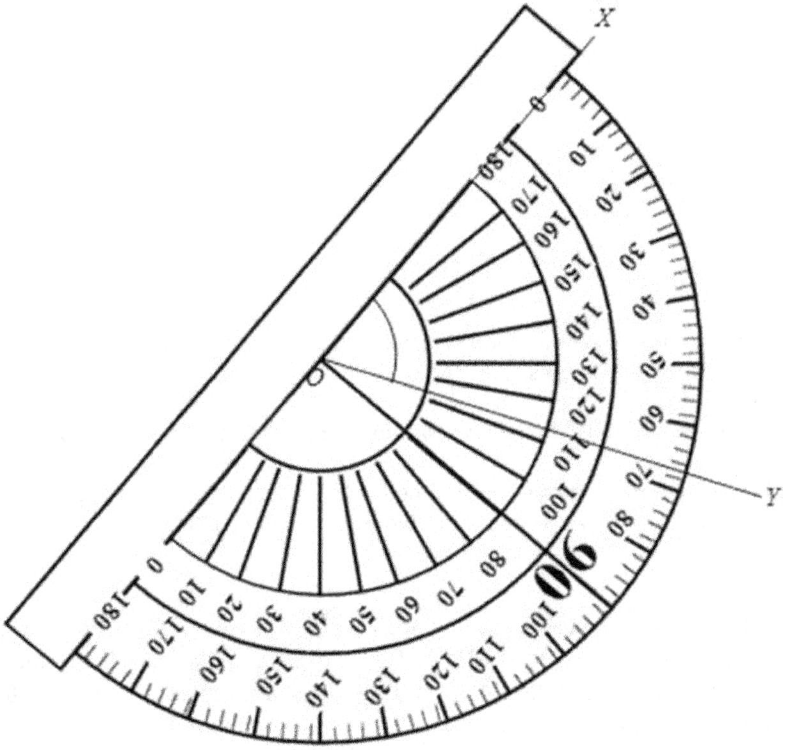

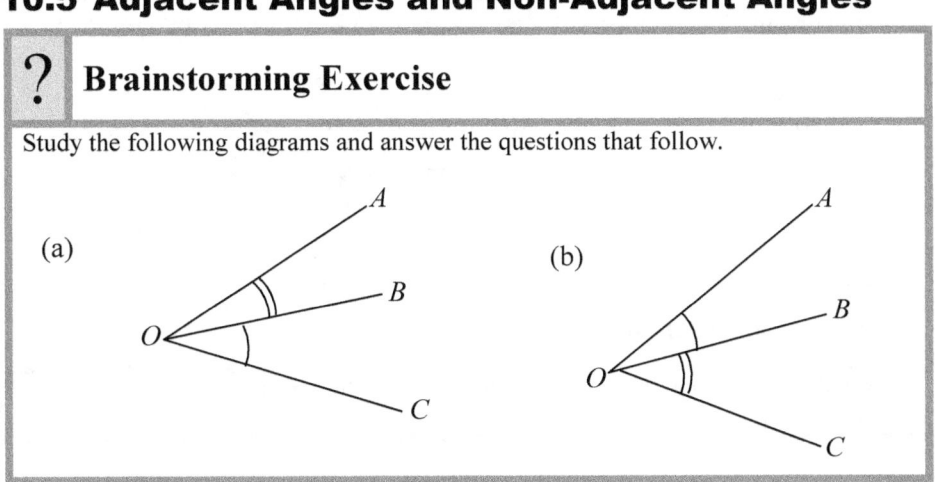

(b)

10.5 Adjacent Angles and Non-Adjacent Angles

? **Brainstorming Exercise**

Study the following diagrams and answer the questions that follow.

(a)

(b)

151

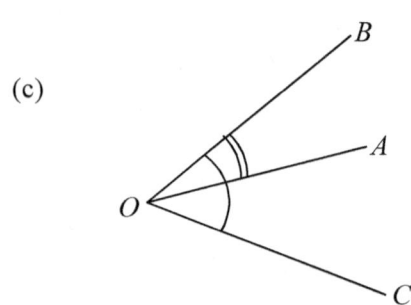

(c)

1. In each of the three diagrams, name the angles marked with arcs.
2. In each of the three diagrams, state the side which is common to the angles marked.
3. In which of the diagrams do the rays forming the marked angles have a common origin?
4. In which of the three angles is the common side between the marked angles?
5. Which of the diagrams fulfill the condition that $\angle AOB + \angle BOC = \angle AOC$?

In the three diagrams, the angles marked are $\angle AOB$ and $\angle BOC$. The side which is common to the angles marked is OB. In figure (a), the rays $[OA)$, $[OB)$ and $[OC)$ have a common origin and $\angle AOB$ and $\angle BOC$ share the common ray $[OB)$ which lies between them and forms one side of each of the angles. Figure (a) satisfies the condition that $\angle AOB + \angle BOC = \angle AOC$.

In such a case $\angle AOB$ and $\angle BOC$ are said to be **adjacent angles**.

In figure (b) and (c) the angles marked are non-adjacent angles. This is because in (b), the rays do not have a common origin. In (c) though the rays have a common origin, the common side (ray) is not between the other two rays.

Therefore, for two angles $\angle AOB$ and $\angle BOC$ to be adjacent,

$$\angle AOB + \angle BOC = \angle AOC.$$

 Example

Given that in the figure below, POQ and QOR are adjacent angles and that $\angle POR$ is equal to $59°$, find the size of x.

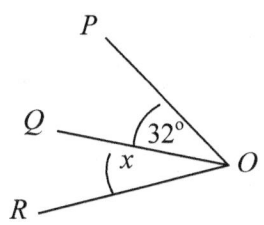

Solution
Since $\angle POQ$ and $\angle QOR$ are adjacent, then
$\angle POQ + \angle QOR = \angle POR$
So to obtain x, we subtract 32° from 59°.
$x = \angle POR - \angle POQ = 59° - 32° = 27°$

10.6 Vertically Opposite Angles, Angles on a Straight Line and Angles at a Point

 ## Review Exercise

In the following diagram. *AB* and *CD* are two lines which intersect. Study the diagram and answer the questions that follow.

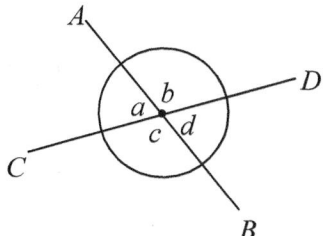

1. What is the special name given to all the angles a, b, c and d?
2. List all the pair of adjacent angles and state their special name.

? Brainstorming Exercise

In the diagram above, which pair of angles is opposite to each other?

Investigative Activity

1. On a cardboard paper draw two intersecting lines and label them as shown in the diagram above.
2. Use a blade or scissors to carve out the angle b.
3. Try to fit this carved out angle b on the angle c.
4. What conclusion do you draw?
5. Deduce the relationship between the angles a and d.
6. State the relationship that exists between the pair of adjacent angles.
7. State the relationship that exists between the angles a, b, c and d.

Adjacent angles on a straight line, sum up to a **half turn** or $180°$.

Angles which sum up to $180°$ are said to be **supplementary**.
So $\angle a + \angle b = 180°$, $\angle b + \angle d = 180°$, and $\angle c + \angle d = 180°$ and $\angle a + \angle c = 180°$.

In the following figure, angles x, y and z are **adjacent angles on a straight line**.

So $\angle x + \angle y + \angle z = 180°$.

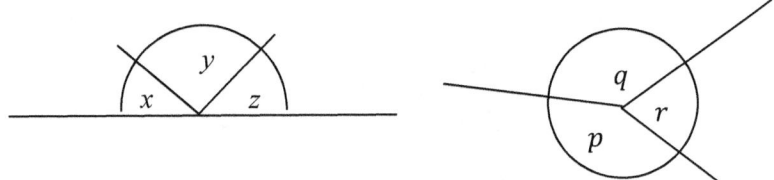

Notice that the angles a, b, c and d (in the diagram under the review exercise) make a **full turn** or a **revolution**. Another name for a full turn is **angles at a point**.

Angles at a point, sum up to a revolution or $360°$.

So $\angle p + \angle q + \angle r = 360°$

Vertically opposite angles are formed when two lines such as AB and CD in the figure above intersect. In the figure above angle a and angle d are pairs of vertically opposite angles. Similarly angle b and angle c are pairs of vertically opposite angles.

Vertically opposite angles are equal. So $\angle a = \angle d$ and $\angle b = \angle c$.

154

Example

In figure (a), (b) and (c) below, find the values of x and y.

(a)

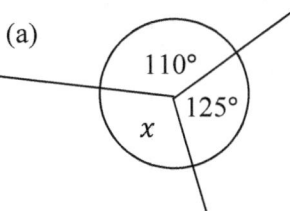

(b)

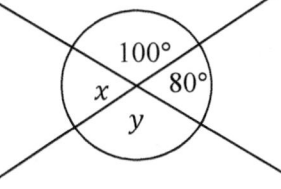

(c)

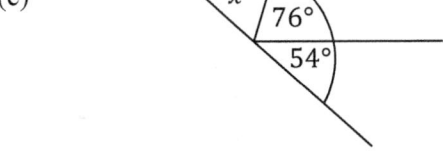

Solution

(a) $x = 360 - (110° + 125°)$ [Angles at a point]
      $x = 125°$

(b)   Vertically opposite angles are equal.
      Therefore, $x = 80°$ and $y = 100°$
      $x = 180° - (76° + 54°)$ [Angles on a straight line]
      $x = 50°$

Exercise 10:2

1. State the relation that exists between the angles marked in the following diagrams.

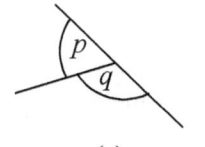

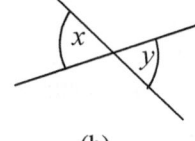

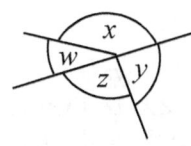

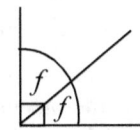

(a)   (b)   (c)   (d)

2. Find the value of the unknowns designated by the letter in each of the following.

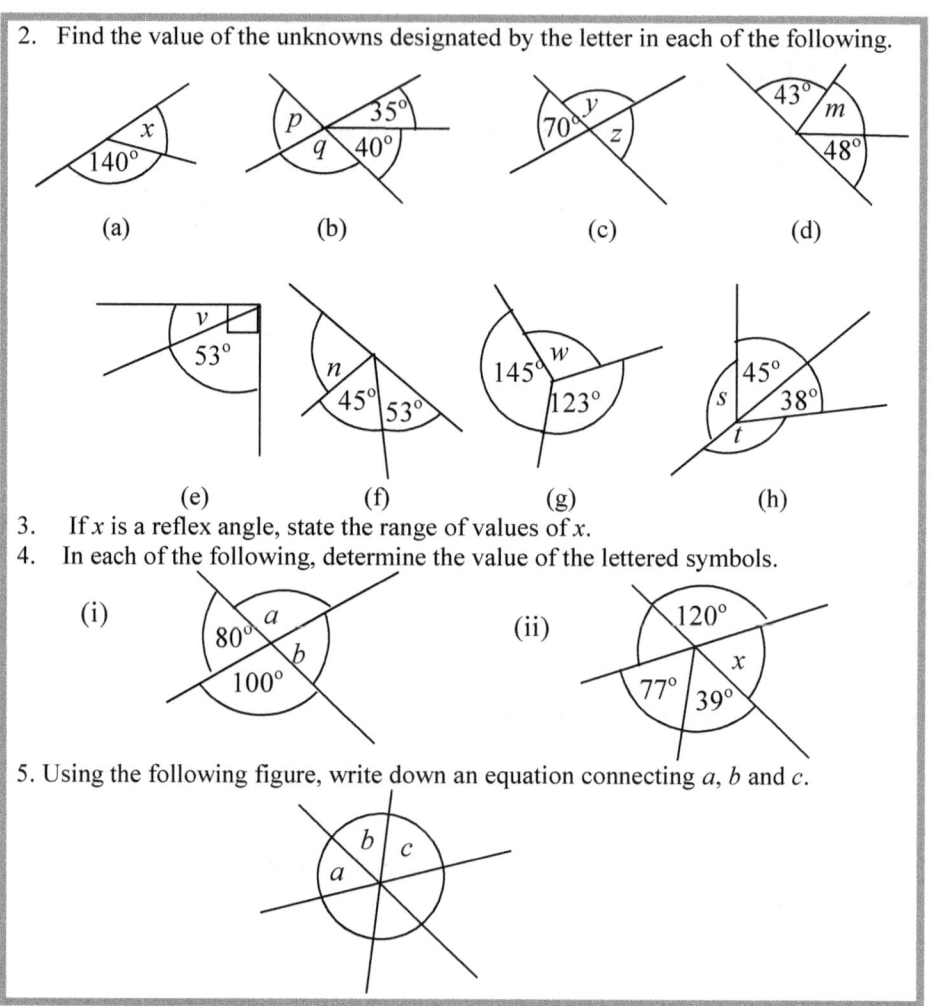

(a)   (b)   (c)   (d)

(e)   (f)   (g)   (h)

3. If x is a reflex angle, state the range of values of x.
4. In each of the following, determine the value of the lettered symbols.

(i)   (ii)

5. Using the following figure, write down an equation connecting a, b and c.

10.7 The Bisector of an Angle

The bisector of an angle is a line which divides the angle into two adjacent angles of the same size.

For instance, in the following figure, $\angle XOZ$ is intentionally drawn to be 60°. If OY is the bisector of $\angle XOZ$, then $\angle XOY = \angle YOZ = 30°$. Confirm this by measuring $\angle XOZ$, $\angle XOY$ and $\angle YOZ$.

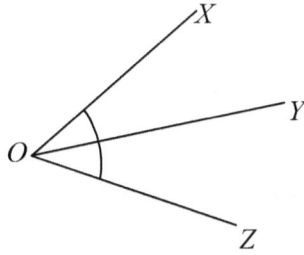

Bisecting a Given Angle *BAC*

Procedure

With center *A* draw 2 arcs of equal radius to cut the adjacent sides to the given angle at two points *C* and *B*.
With centers *C* and *B* draw two arcs of equal radius to intersect at *D*.
Join *A* and *D* with ruler and pencil. *AD* is the bisector of the angle *BAC*.

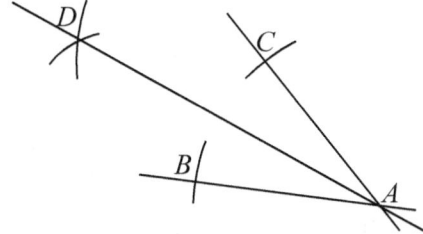

Exercise 10:3

Bisect the following angles and measure the resulting angles.

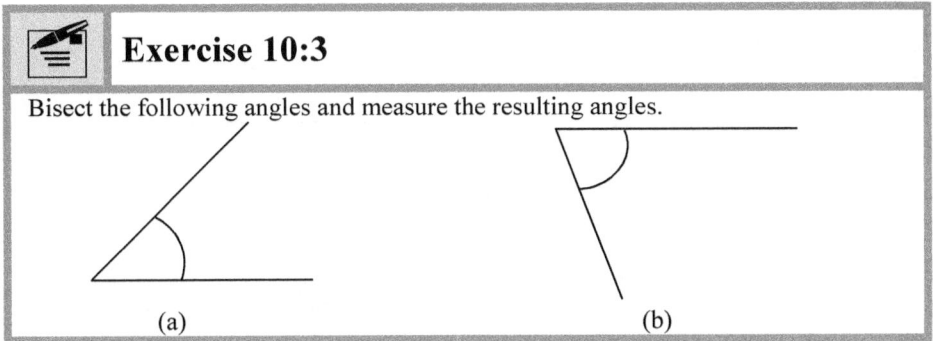

(a)   (b)

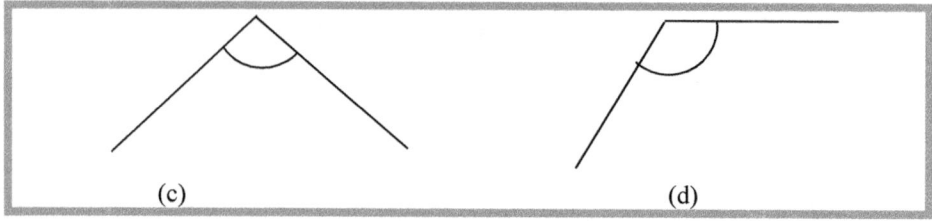

(c)   (d)

✍ Multiple Choice Exercise 10

1. 141° is an example of:
    [A] An acute angle   [B] An obtuse angle
    [C] A reflex angle   [D] An alternate angle
2. An angle whose value lies between 180° and 360° is called:
    [A] A complementary angle   [B] An acute angle
    [C] An obtuse angle   [D] A reflex angle
3. The angle which is not a reflex angle is:
    [A] 316°   [B] 258°   [C] 193°   [D] 117°
4. The angle in figure (a) below is:
    [A] A right-angle   [B] An acute angle
    [C] An obtuse angle   [D] A reflex angle

              (a)   (b)
5. The angle in figure (b) above is:
    [A] straight   [B] acute   [C] right   [D] obtuse
6. In the following and by measurement an angle of 68° is:

              [A]   [B]

              [C]   [D]
7. The measure of angle *A* is 74°. Therefore *A* is:
    [A] right   [B] acute   [C] obtuse   [D] straight
8. The angle in the figure below is certainly:
    [A] obtuse   [B] right   [C] straight   [D] acute

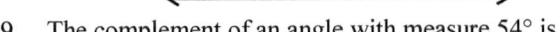

9. The complement of an angle with measure 54° is:
   [A] 144°   [B] 234°   [C] 36°   [D] 126°

10. The value of *x* in (a) below is:
    [A] 42°   [B] 138°   [C] 228°   [D] 132°

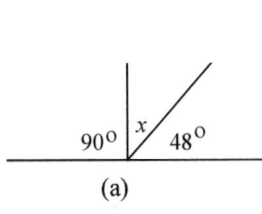

(a)

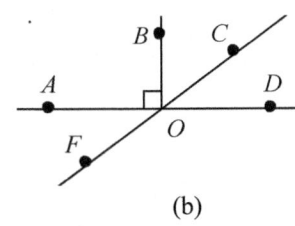

(b)

11. A pair of vertical opposite angles in (b) above are:
    [A] ∠AOF and ∠COD   [B] ∠AOB and ∠DOB

    [C] ∠AOB and ∠COD   [D] ∠BOC and ∠COD

12. A student measured an angle in the intersection of two straight foot paths as 96° shown in the figure below. The value of *s* is:
    [A] 186°   [B] 96°   [C] 192°   [D] 84°

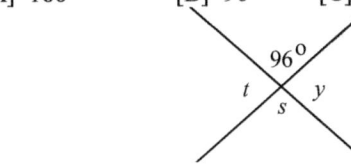

13. A student measured an angle in the intersection of two straight foot paths as 96° shown in the figure above. The value of *t* is:
    [A] 186°   [B] 96°   [C] 192°   [D] 84°

14. The value of *e* in figure (a) below is:
    [A] 50°   [B] 220°   [C] 310°   [D] 130°

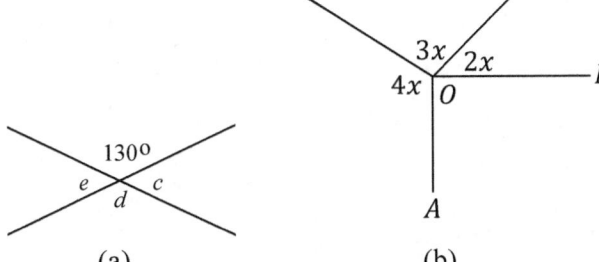

(a)   (b)

15. In figure (b) above, *AO* is perpendicular to *OB*. The value of *x* is:
    [A] 75°   [B] 15°   [C] 22.5°   [D] 30°

16. In the following figure, *AB*, *CD* and *XY* are straight lines intersecting at W. The value of ∠CWX is

[A] 80°   [B] 100°   [C] 120°   [D] 140°

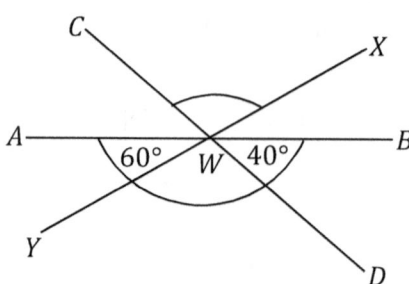

17. The name of the instrument in the figure below is:

[A] a protractor   [B] a pair of dividers
[C] a pair of compasses   [D] a set square

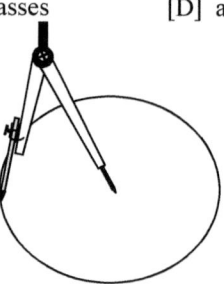

18. The name of the instrument in the figure below is:

[A] a set square   [B] a pair of compasses
[C] a pair of dividers [D] a protractor

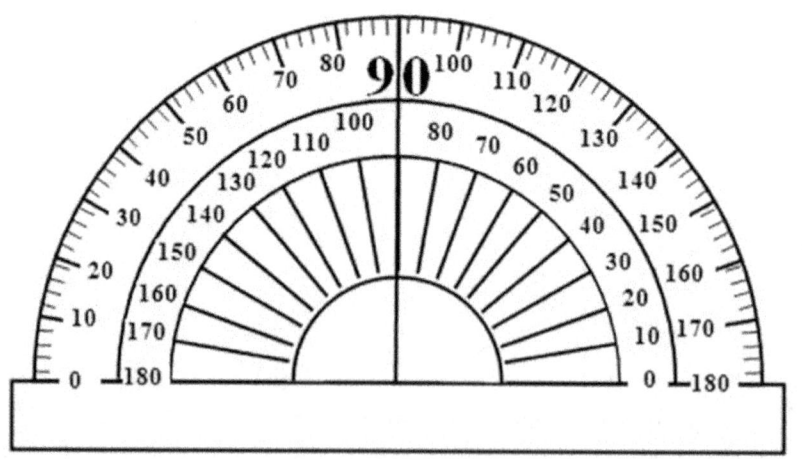

19. A reflex angle is:
    [A] less than 90°   [B] greater than 90°
    [C] greater than 180°   [D] equal to 90°

20. The complement of 47° is:

    [A] 133°   [B] 43°   [C] 313°   [D] −47°

21. If *YN* is the bisector of ∠*XYZ* and $\lambda(\angle XYZ) = 88°$ then, $\lambda(\angle XYN)$ is equal to:

    [A] 176°   [B] 88°   [C] 44°   [D] 22°

22. The following figure shows the construction of:

    [A] A congruent segment   [B] A congruent angle
    [C] A perpendicular bisector   [D] An angle bisector

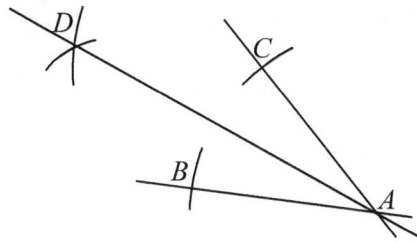

23. The angle bisector of ∠*ABC* is *BD*. If ∠*ABC* = 18°, ∠*ABD* is equal to:

    [A] 30° [B] 36° [C] 18° [D] 9°

Topic 11

PLANE FIGURES

Objectives

At the end of this topic, the learner should be able to:

1. Name, draw and state the properties of the different types of quadrilaterals.
2. State the relationship between the different types of quadrilaterals.
3. Find the perimeter and area of a square, rectangle, a parallelogram, a trapezium, a kite and a rhombus.
4. Identify, draw and name a triangle, using the vertices.
5. Identify, draw, name and classify triangles by measure of their sides and angles.
6. Use the angle sum property of a triangle correctly.
7. Find the perimeter and area of a given triangle when height is given.
8. Identify and construct the height, median and perpendicular bisector of a given a triangle.
9. Name, draw and describe the different parts of a circle.
10. Define and find the circumference of a given circle.
11. Find the area of a circle or disc.
12. Draw concentric circles, circles touching internally externally, disjoint and intersecting circles.

QUADRILATERALS

11.1 Properties of Quadrilaterals

 Group Work

Study the following shapes and answer the questions that follow. Where necessary, first estimate the sides and angles of each shape and then measure it.

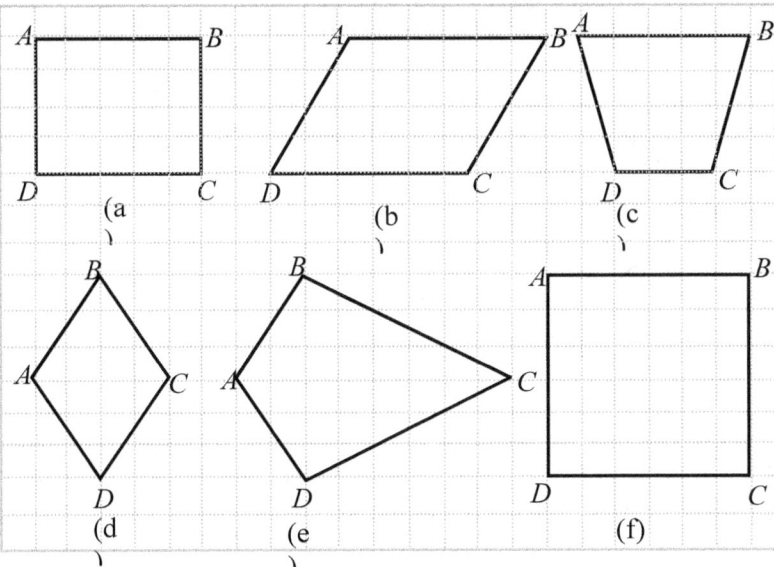

1. How many sides (or edges) does each shape have?
2. How many vertices (corners or angles) does each shape have?
3. Estimate then measure the sides of each shape and record your result on the table below.
4. Which of the shapes have equal sides?
5. In which shapes are opposite sides equal?
6. In which shapes are adjacent sides equal?
7. Which of the shapes have equal angles?
8. In which shapes are opposite angles equal?

	Length of side in cm				Angle in degrees				Name of plane shape
Shape	AB	BC	CD	AD	$\angle A$	$\angle B$	$\angle C$	$\angle D$	
(a)									
(b)									
(c)									
(d)									
(e)									
(f)									

163

9. Estimate then measure the angles of each shape and record your result on the table below.
10. Compare the angles of shape (a) and shape (f) and make a statement concerning them.
11. Draw the diagonals of each shape.
12. How many diagonals can be drawn on each shape?
13. Which of the shapes have diagonals that bisect each other?
14. Which of the shapes have parallel sides?
15. In which shapes are opposite sides parallel?
16. In which shapes do the diagonals meet at right angles?
17. Draw each shape on a cardboard and cut it out using a blade or a scissors.

A **quadrilateral** is a four sided plane figure. Below are the six major types of quadrilaterals. The properties of each quadrilateral deduced in the above group activity are summarized for reference.

1. Trapezium

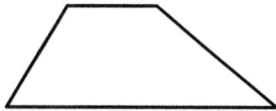

A trapezium is a quadrilateral with two parallel sides.

2. Kite

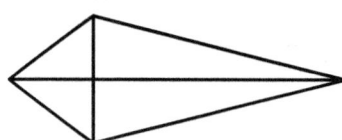

Two pairs of adjacent sides of a kite are equal.
Diagonals of a kite bisect each other at right angles.

3. Parallelogram

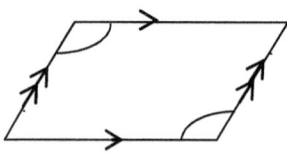

The properties of a parallelogram are:
1. Opposite sides are equal.
2. Opposite sides are parallel.
3. Opposite angles are equal.
4. Adjacent angles are supplementary (sum up to 180°)
5. Diagonals bisect each other.

The Rhombus, the Square and the Rectangle as Parallelograms
The rhombus, the rectangle and the square that follow are all special types of parallelograms. This means that they each have all the properties of the parallelogram together with some other properties.

? Brainstorming Exercise

1. List the properties of a parallelogram that are common to the square, the rhombus and the rectangle.
2. A square is a special rhombus. List the properties which make a square a rhombus.
3. A square is a special rectangle. List the properties which make a square a rectangle.

4. Rhombus

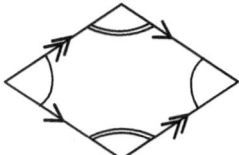

The properties of a rhombus are:
1. All sides are equal.
2. Opposite sides are parallel.
3. Opposite angles are equal.
4. Diagonals bisect each other at right angles.

5. Rectangle

The properties of a rectangle are:
1. Opposite sides are equal.
2. Opposite sides are parallel.
3. Each angle is equal to 90°.
4. Diagonals bisect each other.
5. Diagonals are equal in length.

6. Square

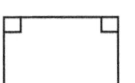

The properties of a square are:
1. All sides are equal. (Different from a rectangle).
2. Opposite sides are parallel.
3. Each angle is equal to 90°. (Different from a rhombus.
4. Diagonals bisect each other at right angles.
5. Diagonals are equal in length.

11.2 Relationship between Quadrilaterals

The relationship between quadrilaterals can be illustrated using the quadrilateral family tree below.

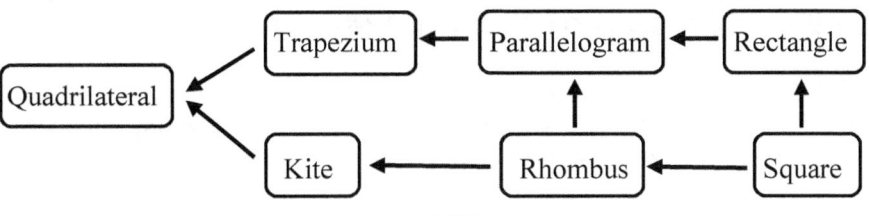

Exercise 11:1

1. Name the quadrilaterals in the following figures (a) to (e).

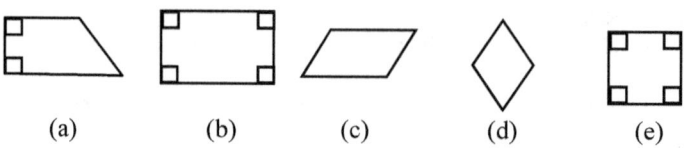

  (a)   (b)   (c)   (d)   (e)

2. In the following table, put 'Y' in the appropriate space if the property holds for the given quadrilateral and 'N' if it does not hold.

	Property	Square	Rectangle	Rhombus	Parallelogram	Kite	Trapezium
a	All the sides are equal						
b	All the diagonals are equal						
c	Diagonals bisect each other						
d	Diagonals are perpendicular						
e	Diagonals bisect opposite angles						
f	Adjacent sides are equal						
g	Opposite sides are equal and parallel						
h	Only two sides are parallel						
i	Adjacent sides are perpendicular						

3. The following figure is a quadrilateral. Find the values of x and y.

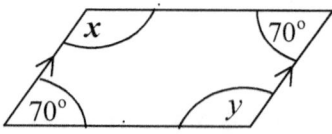

11.3 Mensuration of Quadrilaterals

Perimeter

Perimeter is the length or the distance all round a plane figure. Being a length, perimeter is measured in linear units such as metres (m) and centimetres (cm). Perimeter is often denoted by P.

Area

Area is the amount of surface (or the number of square units) covered by a plane figure. Area is measured in square units such as square metres (m^2), square centimetres (cm^2). The standard unit of area is the square metre. Area is usually denoted by A.

(i) Perimeter and Area of the Rectangle

A rectangle is a quadrilateral with the opposite sides equal and the adjacent sides' perpendicular to each other. The shorter side is called the width or the breadth denoted by w and the longer side is called the length denoted by l.

Length (l)

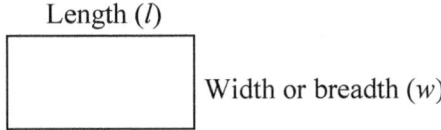

Width or breadth (w)

Investigative Activity

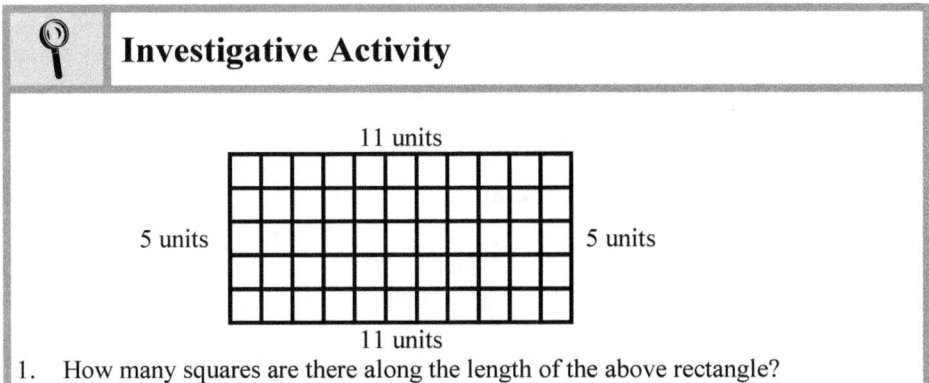

11 units

5 units   5 units

11 units

1. How many squares are there along the length of the above rectangle?
2. How many squares are there along the width of the above rectangle?
3. Add the number of squares on the sides of the rectangle.
4. By counting squares, what is the perimeter the rectangle above?
5. What conclusion do you draw?
6. Find the product of the squares along the length and width of the rectangle.
7. By counting squares, how many squares cover the whole rectangle?
8. What conclusion do you draw?

From the above investigative activity, we see that:

Perimeter = Distance round the rectangle = (5 +11+5+11) units = 32 units

Therefore for a rectangle of length l and width w,

Perimeter$= 2 \times (length + width)$ or $p = 2(l + w)$

From the above investigative activity, we see that by counting squares

Area of rectangle = Number of square units = 55 square units.

Notice that this area could have easily been obtained by multiplying 5 units by 11 units.

Therefore, for a rectangle of length l and width w, the area is given by

$$\text{Area,} = length \times width \text{ or } A = lw$$

 Real life Example

A rectangular floor has sides 4 m by 6 m. Find
(a) the perimeter of the floor   (b) the area of the floor

Solution

(a) $P = 2(l + w) = 2(4 + 6) = 20$ cm   (b) $A = lw = 4(6) = 24$ cm^2

(ii) Perimeter and Area of the Square

? **Brainstorming Exercise**

How can we use the idea of a square as a special rectangle to find the perimeter and the area of a square?

Since a square is a special type of rectangle in which all the sides are equal then. the length and width are equal to say l..

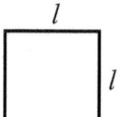

Perimeter of a square, $P = 2(l + l) \Rightarrow P = 4l$

Area of a square, $A = l(l) \Rightarrow A = l^2$

 Example

A square has side 11 cm. Find (a) the perimeter   (b) the area

Solution
(a) $P = 4l = 4(11) = 44$ cm   (b) $A = l^2 = 11^2 = 121$ cm^2

Exercise 11:3

1. A rectangular plot has dimensions as follows.
    (i) 50 m by 35 m   (ii) 90 m by 60 m   (iii) 32 m by 16 m   (iv) 20 m by 15 m
        Find in each case (a) the area   (b) the perimeter.
2. Find the width of a rectangular carpet whose area is 20 cm^2 and whose length is 5 cm.
3. The perimeter of a hall is 33.9 m. Find the length l if the width is 7.6 m.
4. A plank has an area of 1.8 m^2 and a width of 40cm. Find its length in cm.
5. A square room of area 64 m^2 is extended on one side. Find the length of the extension if the new area is 104 m^2.
6. The length of a rectangle is 3 m greater than its breadth. If the area of the rectangle is 180 square metres, calculate its length and its perimeter.
7. Find the length of a rectangle, which has an area of 99 cm^2 and a breadth 9 cm.
8. The area of a rectangle is 40 m^2 and its breadth is 5 m, find its perimeter.
9. Find (a) the area (b) the perimeter, of a square of side (i) 51 cm (ii) 12 cm
        (iii) 15 m   (iv) 30 m
10. Calculate the perimeter of a square whose area is 196 cm^2.
11. The diagonal of a square is 15 cm and the area is 108 cm^2. Calculate
        (a) the length   (b) perimeter of the square.

(iii) Perimeter and Area of the Parallelogram

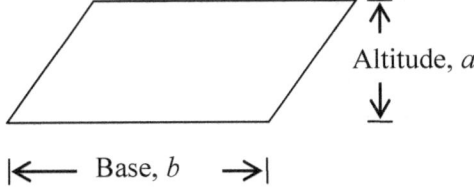

Altitude, a

|← Base, b →|

A parallelogram is a quadrilateral with opposite sides equal and parallel.

 Investigative Activity

(1) Draw a parallelogram on a cardboard and cut it out as in figure (i) below.
(2) Cut out the shaded portion with a blade or scissors and arrange them as in (ii) below.
(3) What do you notice?
(4) How can we use the idea in (3) to find the area of a parallelogram?

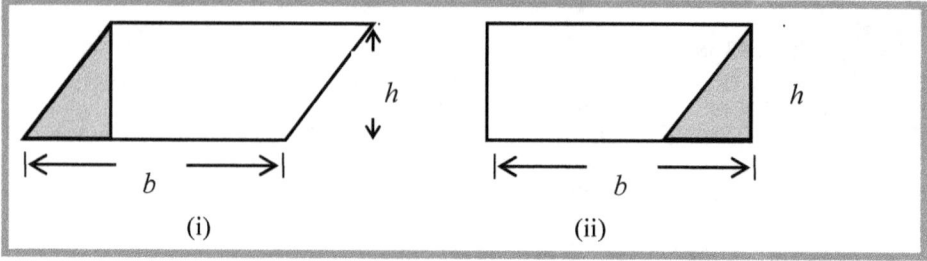

(i)   (ii)

If the shaded portion in figure (i), is cut off and fitted to the right of the same figure, a rectangle whose length and width are *b* and *h* respectively, will be formed as in figure (ii). Therefore, the area of the parallelogram in figure (i) can conveniently be found by calculating the area of the rectangle in figure (ii).

Area of the parallelogram = Area of the rectangle

Area of parallelogram = base × altitude or $A = bh$

 Examples

1. A parallelogram has an altitude of 13 cm and a base 15 cm, calculate its area.

   Solution

   $A = bh = (13 \text{ cm})(15 \text{ cm}) = 195 \text{ cm}^2$

2. Find the altitude of a parallelogram with base 9 cm and area 45 cm².

   Solution

   $h = \dfrac{A}{b} = \dfrac{45}{9} = 5$ m.

(iv) Perimeter and Area of the Rhombus

 Investigative Activity

1. Draw a rhombus with diagonals 6 cm and 10 cm as in figure (a) below [not drawn to scale].
2. Measure the length of the sides of the rhombus.
3. Find the perimeter of the rhombus.
4. Use a blade or scissors to cut it along the diagonals.
5. Arrange the pieces as in figure (b) below.

6. What is the resulting figure?
7. What are the length and width of the resulting figure?
8. Find the area of the resulting figure.
9. Deduce the area of a rhombus?

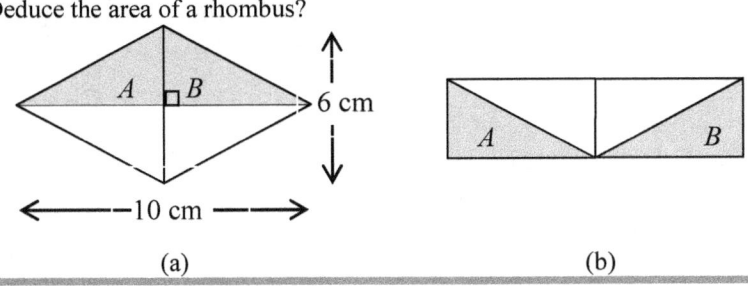

(a)   (b)

From the above investigation, it can be seen that for a rhombus whose sides are of length l and whose diagonals are x and y (figure (a) below).

$$\text{Area } A, \text{ of rhombus} = \text{Area of rectangle} = \text{length} \times \text{width} = \frac{1}{2}xy$$

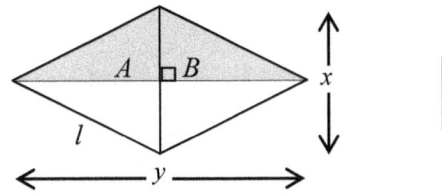   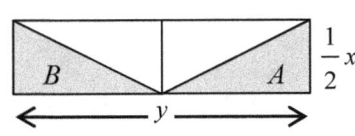

Area of a rhombus $= \frac{1}{2} \times$ product of the diagonals

Perimeter of a rhombus $= 4 \times$ length of side $= 4l$

 Example

Find the area and perimeter of a rhombus whose diagonals are 12 cm by 16 cm and whose sides are 10 cm.

Solution

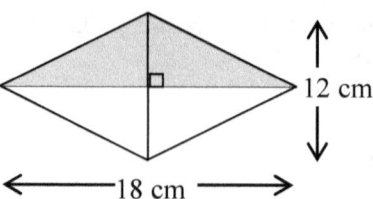

Area $= \frac{1}{2}xy = \frac{1}{2}(12)(16) = 96$ cm^2

Perimeter $= 4 \times 10 = 40$ cm.

(v) Perimeter and Area of the Trapezium

 Investigative Activity

(1) Draw a parallelogram with the parallel sides 6 cm and altitude 3 cm as in figure below.
(2) Calculate the area of this parallelogram.
(3) From one vertex, measure 4 cm and mark the point. From the opposite vertex measure again 4 cm and mark the point.
(4) Draw a line to connect the two points you have marked.
(5) Use a blade or scissors to cut the parallelogram along the line you have drawn.
(6) What are the resulting figures?
(7) Flip one of the figures and try to fit it on top of the other. Do they fit exactly?
(8) Deduce the area of one of your figures from the area of the parallelogram you found in (2) above.

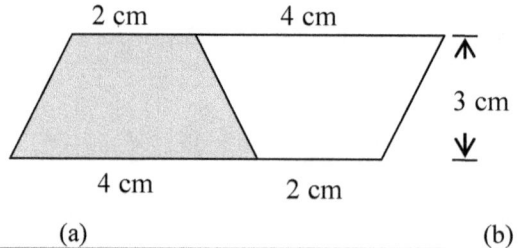

(a)   (b)

From the above investigation, we see that a trapezium with parallel sides of length a and b and altitude h, can be visualized as half of a parallelogram whose bases are of length $(a + b)$ and whose altitude is h.

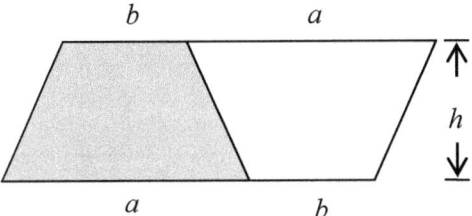

Area of trapezium $= \frac{1}{2} \times$ Area of parallelogram $= \frac{1}{2}$ base $\times$ altitude

Area of trapezium, $A = \frac{1}{2}(a + b)h$

Area of trapezium $= \frac{1}{2} \times$ sum of parallel sides $\times$ altitude

 Example

Find the area of a trapezium whose bases are 6 cm and 4 cm with the distance between parallel sides 5 cm.

Solution

Area of trapezium $= \dfrac{1}{2} \times$ sum of parallel sides $\times$ altitude

Area of trapezium $= \frac{1}{2} \times (a + b)h = \frac{1}{2}(6 + 4)5 = 25$ cm^2.

 Exercise 11:4

1. Find the area and perimeter of the parallelogram below

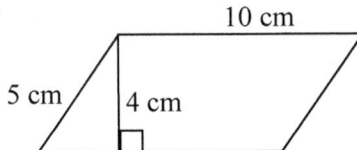

2. A flower bed in the form of a parallelogram has opposite sides of length 10 cm. The distance between these opposite sites is 6 cm. Calculate the area of the flower bed.
3. The diagonals of a rhombus are 12 cm and 16 cm. Calculate its area.
4. Given that the area of the parallelogram in figure (a) below is 54 cm^2. Find the altitude of the parallelogram.

173

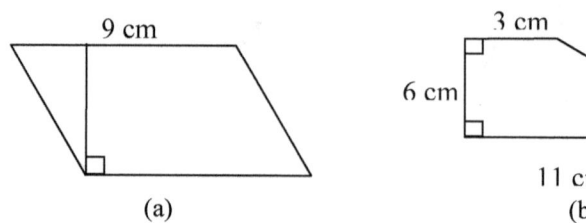

(a) (b)

5. In Figure (b) above, calculate the area of the trapezium.
6. In the figure below, calculate
    (a) The perimeter of the trapezium.   (b) The area of the trapezium.

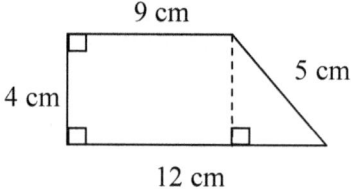

7. Calculate the area of the trapezium in the figure below.

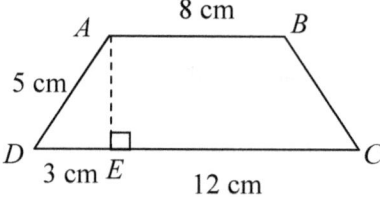

8. The parallel sides of a trapezium are 13 cm and 9 cm each. If the distance between them is 12 cm, find the area of the trapezium.
9. Find the length of one of the parallel sides of a trapezium given that the height is 8 cm, the length of the other parallel side is 10 cm and the area is 124 cm^2.
10. The side of a rhombus is 30 cm and the diagonals are 24 cm and 36 cm. Find
    (a) The perimeter of the rhombus   (b) The area of the rhombus.
11. In the figure below, the area of trapezium *ABCE* is 54 cm^2. Find the value of *BC*.

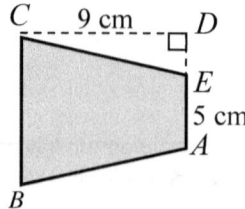

12. A decoration in front of a hall is made up of four diamond-like flower beds with equal sides and diagonals of lengths 2 m by 3 m. Calculate the total area covered by the flower bed.

174

TRIANGLES

11.4 Properties of Triangles

A triangle is a three sided plane figure.

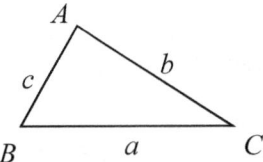

In labeling a triangle, uppercase or capital letters are used to label the vertices (corners) as shown in the figure above. The side opposite a vertex is labeled using its corresponding lowercase or small letter.

For instance the side opposite the vertex *A* is labeled using the letter *a*. The triangle above is referred to as triangle *ABC* or triangle *BCA* or triangle *BAC*. The order of the letters does not matter however it sounds better using the letters in their alphabetical order.

Triangles are classified in two ways as follows.

(a) *Naming Triangles by the Measures of their Angles*

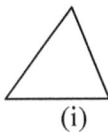   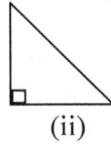   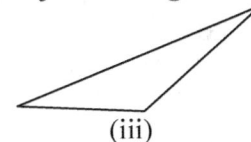

(i)   (ii)   (iii)

(1) **An Acute angle triangle** (figure (i) above) has all the angles each less than 90°.
(2) A **right-angled triangle** (figure (ii) above) has one of its angles equal to 90°.
(3) **Obtuse angle triangle** (figure (iii) above) has one of its angles between 90° and 180°.

(b) *Naming Triangles by the Measures of their Sides*

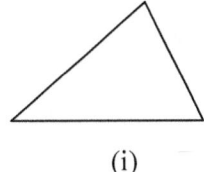   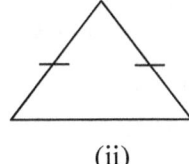   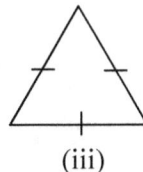

(i)   (ii)   (iii)

(1) **A Scalene triangle** (figure (i) above) has no sides equal.

(2) **An Isosceles triangle** (figure (ii) above) has two sides equal.

(3) **An Equilateral triangle** (figure (i) above) has all the sides equal.

 Exercise 11:5

1. Classify the triangles below by the measures of their sides.

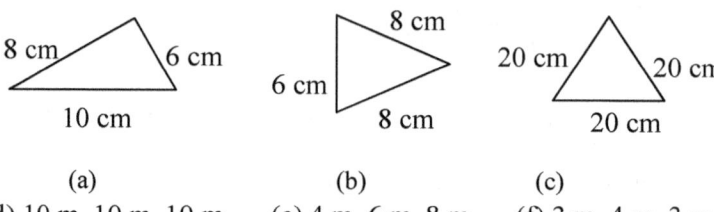

| (a) | (b) | (c) |

(d) 10 m, 10 m, 10 m   (e) 4 m, 6 m, 8 m   (f) 3 m, 4 m, 3 m

2. Classify the triangles below by the measures of their angles.

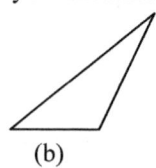 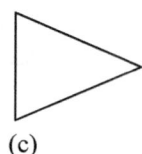

(a)   (b)   (c)

3. In the following, two angles of a triangle are given. Name the type of triangle.
    (a) 50°, 50°   (b) 30°, 60°   (c) 24°, 85°

4. Draw the following:
    (a) an acute angle triangle *ABC*.   (b) a right-angled triangle *XYZ*.
    (c) an obtuse angle triangle *PQR*.   (d) a scalene triangle *RST*.
    (e) an isosceles triangle *UVW*.   (f) an equilateral triangle *LMN*.

5. Answer true or false
    (a) an isosceles triangle is always an acute angle triangle.
    (b) an equilateral triangle can sometimes be a right angled triangle
    (c) an equilateral triangle is always an acute angle triangle.
    (d) an isosceles triangle can sometimes be a right angled triangle.
    (e) a scalene triangle can sometimes be a right angled triangle.
    (f) a right-angled triangle can never be an equilateral triangle.

6. Classify the following triangles by the measures of their sides.
    (a) A triangle with sides 8 cm, 6 cm, 8 cm.
    (b) A triangle with sides 2 cm, 3 cm, 4 cm.
    (c) A triangle with sides 3 cm, 4 cm, 5 cm.
    (d) A triangle with sides 9 cm, 9 cm, 9 cm.

7. Classify the following triangles by the measures of their angles.
    (a) A triangle with angles 40°, 40°, 100°.
    (b) A triangle with angles 30°, 60°, 90°.
    (c) A triangle with angles 20°, 75°, 85°.

11.5 Sum of Interior Angles of a Triangle

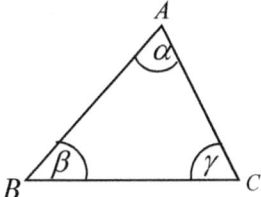

In the figure above, the angles α, β and γ shown are called the **interior angles** of the triangle, because they are inside the triangle.

 Investigative Activity

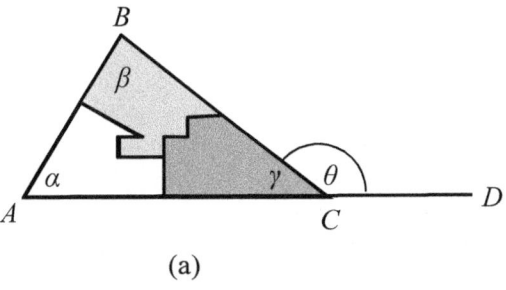

(a)

(1) Draw a triangle ABC (any size) and produce AC to D as shown in figure (a) above.
(2) Cut out the vertices A and B of the triangle ABC and arrange at C as in figure (b) below.

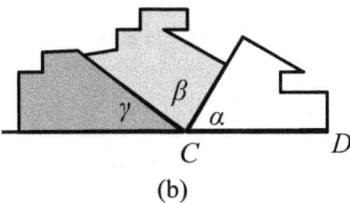

(b)

(3) Do they fit exactly onto the angle θ the exterior angle of the triangle?
(4) Do the three angles α, β and γ lie and fit exactly on the straight line AD?
(5) Use your answer in (4) to deduce the numerical value of $\alpha + \beta + \gamma$?
(6) Make a general statement concerning the interior angles of a triangle.

Since the angles α and β fit into the angle θ and all the three angles lie on a straight line, the above investigation reveals that

(1) The exterior angle of a triangle is equal to the sum of the two interior opposite angles.

(2) The sum of the interior angles of a triangle is 180° (or two right angles). In other word, $\alpha + \beta + \gamma = 180°$

Exercise 11:6

Find the value of the lettered angles in figure (a) to (d) below.

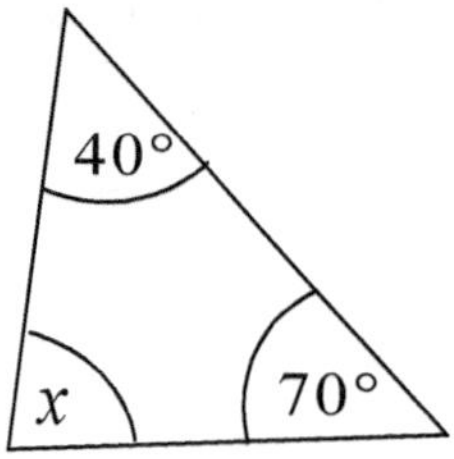

(a)

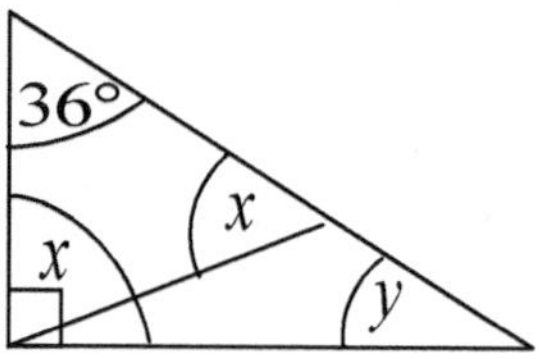

(b)

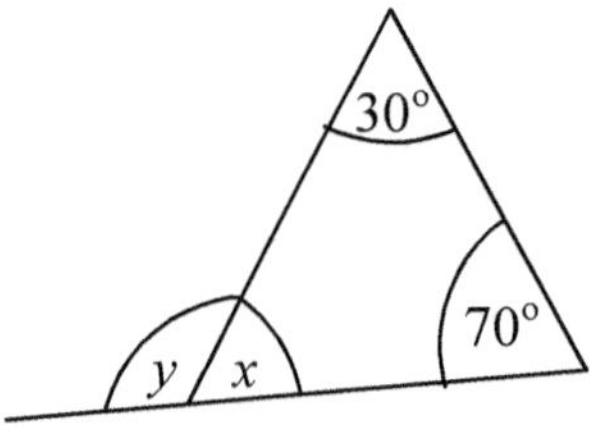

(c)

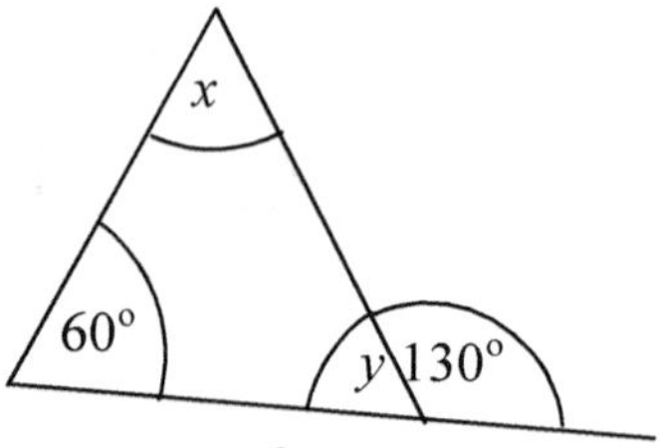

(c)

11.6 Mensuration of Triangles

 Investigative Activity

(1) Draw a parallelogram with the parallel sides 6 cm and altitude 3 cm as in figure below.

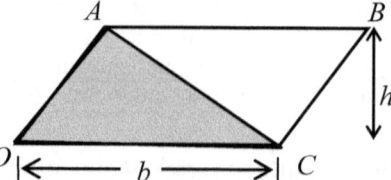

(2) Calculate the area of this parallelogram.

(3) Draw the diagonal from *A* to *C*.
(4) Use a blade or scissors to cut the parallelogram along the *AC*.
(5) What are the resulting figures?
(6) Flip one of the figures and try to fit it on top of the other. Do they fit exactly?
(7) Deduce the area of one of your figures from the area of the parallelogram you found in (2) above.
(8) Measure the sides of one of the resulting figures.
(9) Find the perimeter of one of the resulting figures.

From the above investigation, we see that a triangle with base *b* and altitude *h* can be visualized as half of a parallelogram whose bases are of length *b* and whose altitude is *h*. Therefore

$$\text{Area } A, \text{of triangle} = \frac{1}{2} \times \text{Area of parallelogram} = \frac{1}{2}bh$$

Perimeter *P*, of triangle = Sum of the sides = $a + b + c$

Real life Example

A triangular lawn has a base of 4 m and a height of 8 m. What is its area?

Solution

$$A = \frac{1}{2}bh = \frac{1}{2}(8)(4) = 16 \text{ cm}^2$$

Exercise 11:7

1. Calculate the area of a triangle whose base is 12 cm and whose height is 5 cm.
2. The area of a triangle is 36 cm². Given that the base is 12 cm, find the height of the triangle.
3. Calculate the area of an isosceles triangle whose base is 10 cm and whose altitude is 6 cm.
4. The perimeter of an isosceles triangle is 16 cm and the length of each of the two equal sides is 5 cm. Find
   (a) The length of the third side   (b) The altitude of the triangle
   (c) Its area given that the height is 4 cm.

11.7 Height, Median and Mediator of a Side of a Triangle

The **altitude** (or **height**) *AH*, of a triangle is the perpendicular distance from a vertex to the straight line which contains points on the opposite side of the triangle.

The **perpendicular bisector** or **mediator** *XP* of a side of a triangle is a line that is perpendicular to this side and passes through the midpoint of the side.

The **median** *AM* of a triangle is a line from the midpoint of a side of the triangle to the opposite vertex.

In the following three figures, *AH*, *AM* and *XP* are the altitude or height, the median and perpendicular bisector respectively of the triangle with respect to the side *BC* of the triangle.

1.

2.

3.

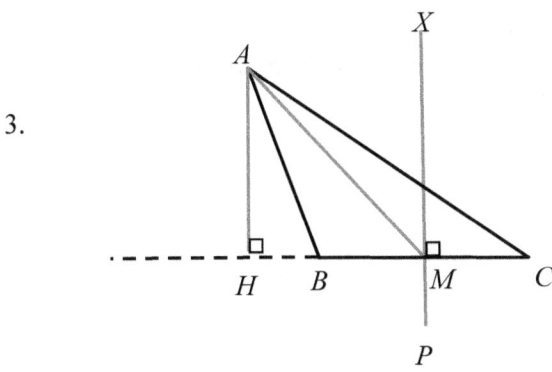

Constructing the Perpendicular from a Vertex to the Opposite Side of a Triangle

We can use the procedure in the previous example to construct the perpendicular from a vertex to the opposite side of a triangle and hence determine the altitude or height of the triangle with respect to the side.

To construct the median of a side of a triangle, construct the perpendicular bisector of the side of the triangle and connect their point of intersection to the vertex of the triangle.

✎ Example

Construct the perpendicular from the vertex A to the base BC of the following triangle.

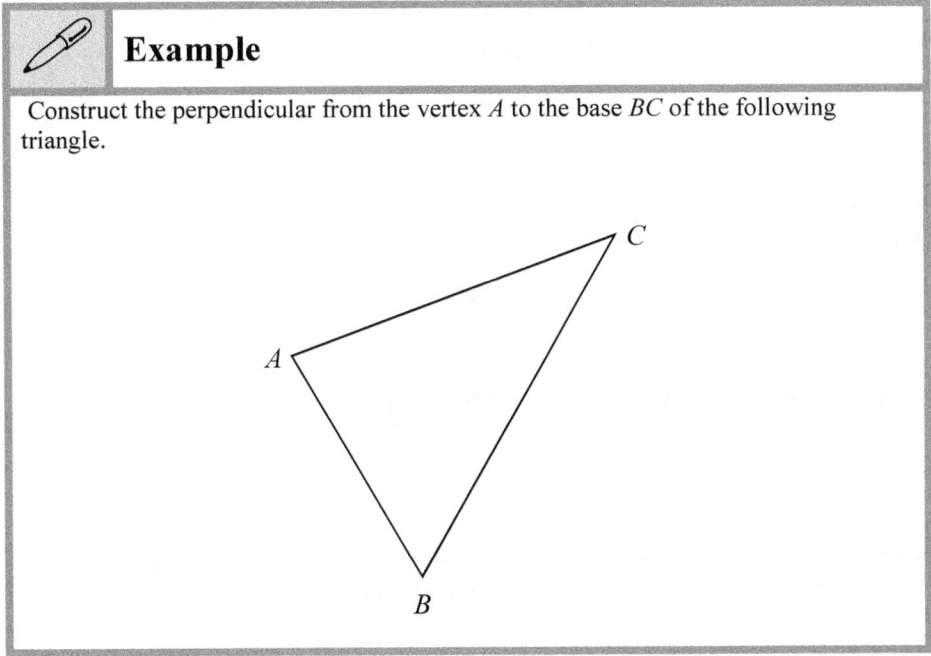

Procedure

(i)   With center *A* draw 2 arcs of equal radii to cut *BC* at two points *P* and *Q*.
(ii) With centers *P* and *Q* draw two arcs of equal radii to intersect on the opposite side of *A*. Name this point of intersection *D*.
(iii) Now join *AD*. *AD* is perpendicular to *BC*.
(iv) Name the point of intersection of *BC* and *AD*, *N*.

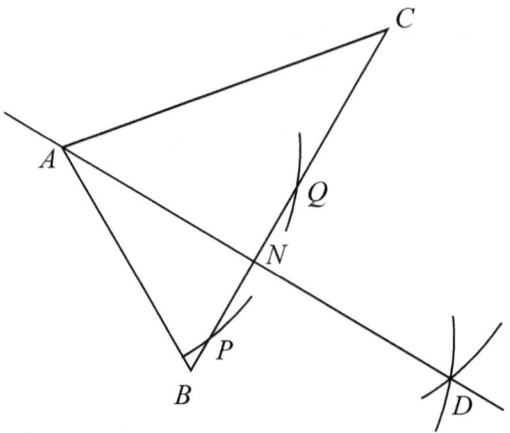

(i)   *AN* is the altitude or height of the triangle *ABC*.

Constructing the Orthocenter of a Triangle

The orthocenter of a triangle is the point of intersection of the perpendicular bisectors of the sides of a given triangle.

 Exercise 11:7

Construct the orthocenter of the triangle above.

Constructing the Median of a Triangle

 Example

Construct the perpendicular bisector of the side *BC* of triangle in the previous example. Hence, construct the median to this side of the triangle.

Construct the perpendicular bisector of the side *BC* of triangle in the previous example. Hence, construct the median to this side of the triangle.

Procedure
(i)   With centers *B* and *C* draw two arcs of equal radii to intersect on the opposite sides of *BC*. Name these points of intersection *X* and *Y*.
(ii) Now join *XY*. *XY* is the perpendicular bisector of *BC*.
(iv) Name the point of intersection of *BC* and *XY*, *M*.
(v)   Draw a straight line to pass through *A* and *M*. *AM* is the median of the triangle *ABC*.

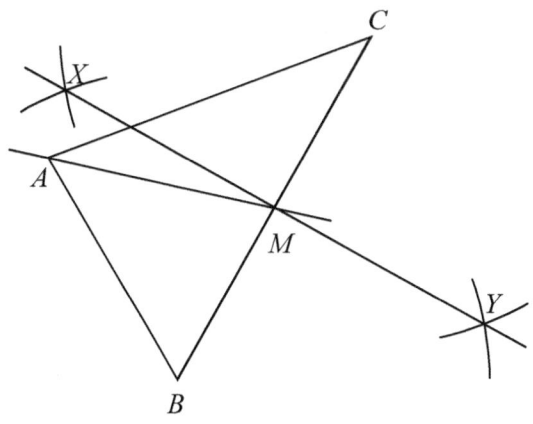

Constructing a Triangle ABC with Sides of Given Length

 Example

Construct a triangle *ABC* with sides of length *AB* = 5 cm, *BC* = 9cm and *AC* = 13 cm.

Procedure
(i)   Following the steps in 1, construct the line *AB* of length 13 cm.
(ii) With compass, measure 9 cm and with center *B*, draw an arc on one side of the line.
(iii) With compass, measure 5 cm and with center *A*, draw another arc to cut the first. Mark their point of intersection *C*.
(iv) Using ruler and pencil join *AC* and *BC*

183

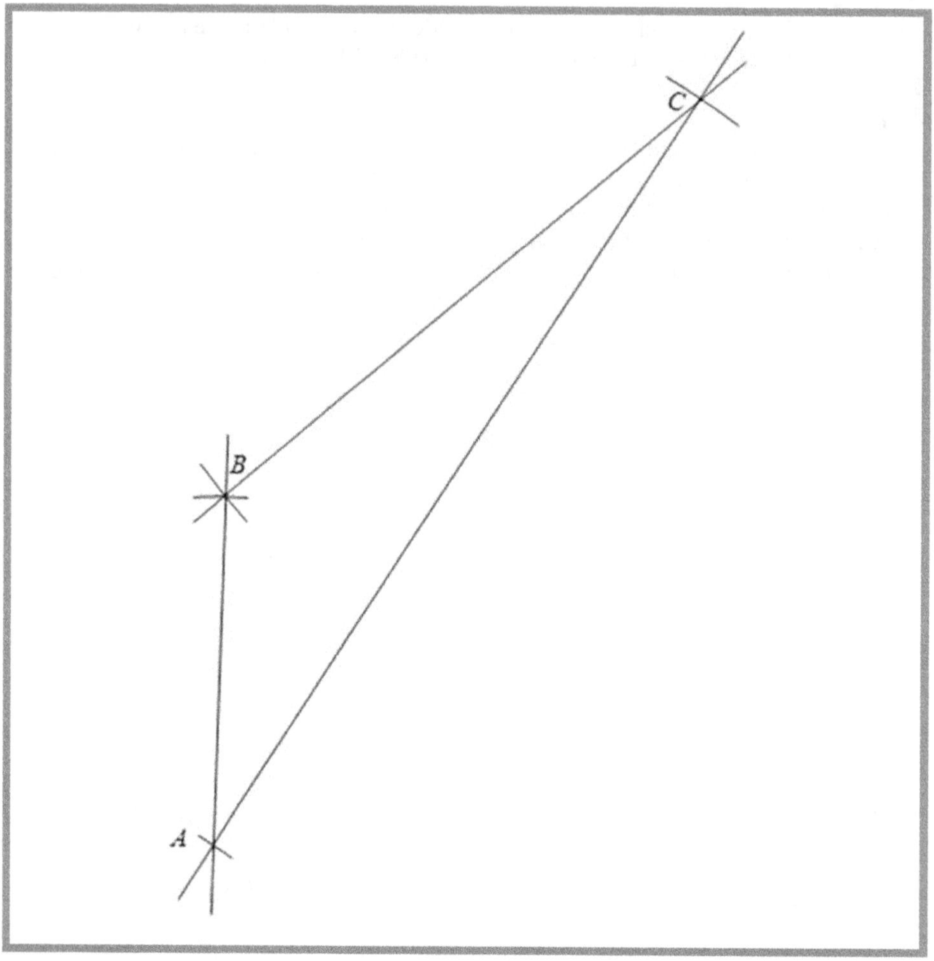

Note!! Construction lines do not end at the vertices *A*, *B*, and *C*.

Exercise 11:8

In this exercise use only a ruler, a pair of compasses and a pencil, and show all construction lines.

1. Construct a triangle *LMN* with sides *LM* = 7 cm, *LN* = 8 cm and *MN* = 6 cm.
2. (i) Draw a line *OA* of length 8 cm.
   (ii) Construct a line *OB* of length 6 cm perpendicular to *OA*.
   (iii) Measure *AB*.
   (iv) Bisect *AB* (by construction only) and mark the midpoint *M* of *AB*.
   (v) Measure *MO*.
3. (a) Draw a line segment, *PQ*, 7 cm long in the middle of a new page
   (b) Bisect the line *PQ* and label the mid-point, *X*.

4. (a) Draw a line *PQ* of length 8 cm.
   (b) Construct a line *OQ* of length 6 cm perpendicular to *PQ*.
   (c) Draw and measure the line *OP*.
   (d) Construct the perpendicular bisector of *OP* and label its foot *M*.

CIRCLES

A circle is a plane figure bounded by points, which are equidistant from a fixed point called the centre of the circle.

11.8 Vocabulary Related to Circles

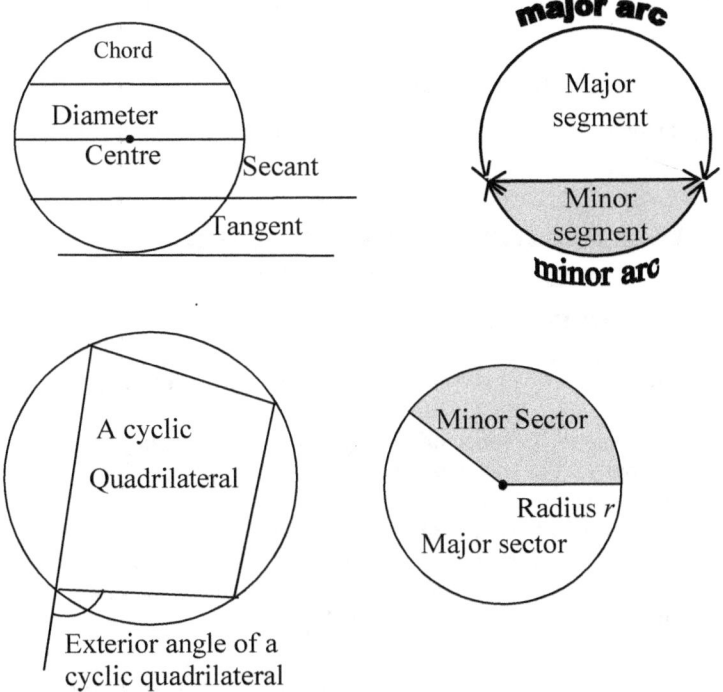

The **circumference** of a circle is the distance round the circle. It is usually denoted by *C*.

The **chord** is a straight line drawn from one point on the circumference to another point on the circumference.

The **diameter** is a straight line drawn from one point on the circle passing through the centre of the circle to another point on the circumference. The diameter is the longest chord.

The **radius** is a straight line drawn from the centre of a circle to a point on the circumference of the circle.

An **arc** is part of the circumference. The longer of the arcs is called the **major arc** while the shorter is called the **minor arc.**

A **sector** is a portion of a circle bounded by two radii and an arc of the circle. The larger of the sectors is called the **major sector** while the smaller is called the **minor sector.**

A **segment** is a portion of a circle bounded by a chord of the circle and an arc of the circle. The smaller segment is called the **minor segment** while the larger segment is called the **major segment.**

When a chord is produced to extend outside the circle, it is called a **secant.**

A straight line which is drawn to pass through one and only one point of a circle is called a **tangent.**

Concentric circles are circles with a unique centre.

A **cyclic quadrilateral** is a quadrilateral which is inscribed in a circle.

11.9 Relative Position of two Circles

> ? **Brainstorming Exercise**
>
> Draw pair of circles as described and chose from the following the phrase that best fit your pair of circles.
>   (i) disjoint circles   (ii) concentric circles   (iii) intersecting circles
>   (iv) touch internally   (v) touch externally
> (1)   A pair of circles that are apart.
> (2)   A pair of circles with the same center but different radii.
> (3)   A pair of circles that cut each other at two different points.
> (4)   A pair of circles that touch in such a way that one is inside the other.
> (5)   A pair of circles that touch but none is inside the other.

1. **Disjoint circles** are circles that neither touch nor intersect each other as in figure (e) below.
2. **Concentric circles** are circles which have the same centre as in figure (d) below.
3. **Intersecting circles** that touch at one point or cut each other at two distinct points as in figure (a), (b) and (c) below.
4. When circles touch each other at one point only and one is inside the other as in figure (a) below we say the circles **touch internally**.

5. When circles touch each other at one point only but none is inside the other as in figure (b) below we say the circles **touch externally**.

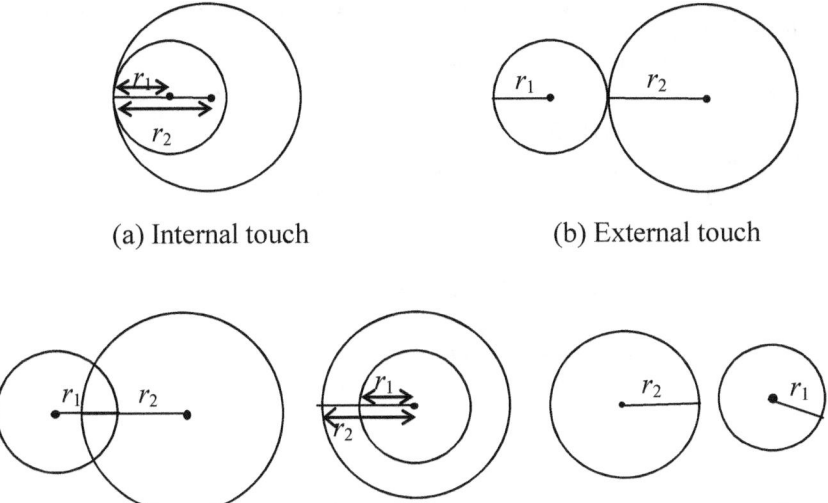

(a) Internal touch   (b) External touch

(c) Intersecting circles   (d) Concentric circles   (e) Disjoint circles

---

 Investigative Activity

Investigate and match up the phrases in group A with those in group B to complete each statement.

GROUP A
(a) When two circles touch internally,_____
(b) When two circles touch externally, _____
(c) When two circles are disjoint,_____
(d) When two circles intersect at two different positions,_____
(e) When two circles are concentric,_____

GROUP B
1. The distance between their centers is zero.
2. The distance between their centers is equal to the difference between their radii.
3. The distance between their centers is greater than the sum of their radii.
4. The distance between their centers is equal to the sum of their radii.
5. The distance between their centers is less than the sum of their radii but greater than the difference between their radii.

Exercise 11:9

1. In the following figure,
    (i) Name the set of points on the line
        (a) *AC*   (b) *OC*   (c) *FD*   (d) *ABC*
        (e) *ACX*   (f) *EY*   (g) *AEC*
    (ii) Name the set of points in the region.
        (a) *DEFD*   (b) *CODC*   (c) *ABCA*

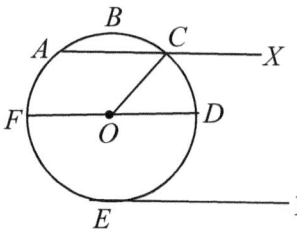

2. Figure (a) below shows a circle with centre *O* and points *A*, *B* and *C* on the circle. *ON* is perpendicular to *AB*. Give a name for each of the following:
    (a) The set of points on *ACB*.   (b) The set of points on *OA*.
    (c) The triangle *OAB*.   (d) The set of points on *AB*.
    (e) The set of points in the region *OACBO*.

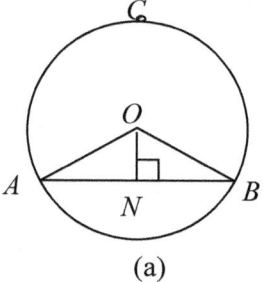   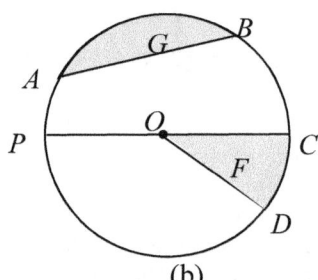

         (a)   (b)

3. With respect to Figure (b) above, where *O* is the centre of the circle, name
    (a) the shaded region *G*   (b) the curve *CD*   (c) the line *OD*
    (d) the line *PC*   (e) the line *AB*   (f) the shaded region *F.*

11.10   Mensuration of the Circle

Relationship between circumference and diameter [*Pi (π)*]

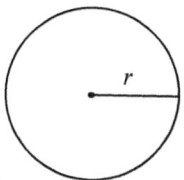

 Investigative Activity

Requirements
In this activity you need tins of different sizes, a sticky tape, razor blade and a ruler.

Procedure

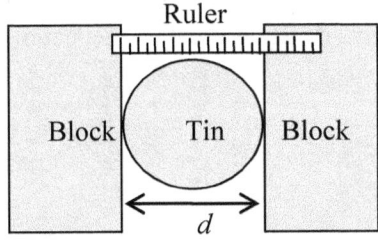

1. Rap a sticky tape around each tin in such a way that it surrounds the tin and overlaps.
2. Cut the sticky tape from each tin with a razor and peel it off.
3. Paste the peeled off tape on a flat board and measured it with a ruler.
4. Measure the diameter of the tin by placing it in-between two blocks and measuring the distance between them as shown below.
5. Record your result on the following table.

Tin	Circumference, C (cm)	Diameter, d (cm)	$\dfrac{C}{d}$
Tin 1			
Tin 2			
Tin 3			
Tin 4			
Tin 5			
Tin 6			
Tin 7			

6. Divide the circumference in each case by the diameter to obtain $\frac{C}{d}$.
7. What do you notice?

The following table shows the result of one such experiment.

Tin	Circumference, C (cm)	Diameter, d (cm)	$\frac{C}{d}$
Tin 1	22.5	7	3.2
Tin 2	28.6	9.2	3.1
Tin 3	12.7	4	3.2
Tin 4	20	6.4	3.1
Tin 5	40	12.6	3.2
Tin 6	9.1	3	3.0
Tin 7	15.4	5	3.1

The results clearly show that the ratio $\frac{C}{d}$ is almost the same for all the tins.

More accurate measurements show that this ratio is approximately 3.142, sometimes approximated to $\frac{22}{7}$. This constant is called pie, denoted by π. Thus,

$$\pi = 3.142 \approx \frac{22}{7}$$

Since $\pi = \frac{C}{d}$ it means that $C = \pi d$

But $d = 2r$, so $C = 2\pi r$

Examples

1. Calculate the circumference of a circle whose radius is 21 cm, taking $\pi = \frac{22}{7}$.

   Solution

   $C = 2\pi r = 2\left(\frac{22}{7}\right)(21) = 132$ cm

Area of a Circle

 Investigative Activity

1. Draw a circle of radius 3 cm.
2. Divide the circle into 18 equal sectors as shown in figure (a).
3. With a blade or scissors cut out the sectors.
4. Arranged the sectors as shown in figure (b).
5. What figure does the shape resemble?
6. How do we find the area of the shape this shape resembles?
7. What is the length of each of the bases compared to the circumference of the circle?

8. What is the altitude of the shape compared to the radius of the circle?

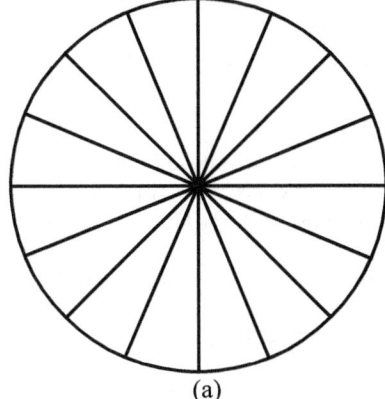

(a)

9. Use your ideas to find the area of the shape.

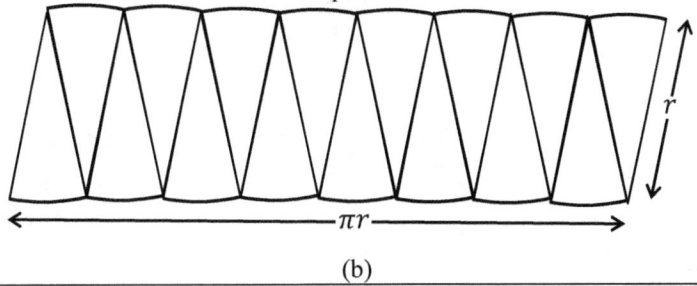

(b)

By cutting a circle into a number of sectors and arranging the sectors as in figure (b) above, a shape similar to a parallelogram will be obtained. If the sectors, are many and smaller the figure can be approximated to a parallelogram whose base is πr (half of the circumference) and whose altitude is r.

Therefore,

Area of circle $\approx$ area of the parallelogram $=$ base $\times$ height $= \pi r(r)$

So Area A, of circle $= \pi r^2$

Example

Calculate the area of a circle whose radius is 14 cm. Take $\pi = \frac{22}{7}$.

Solution

$$A = \pi r^2 = \frac{22}{7}(14)^2 = 616 \text{ cm}^2$$

 Exercise 11:10

Where necessary in this exercise take $\pi = \frac{22}{7}$.

1. Find the area of a circle whose radius is 3.5 cm.
2. Calculate the maximum area a goat will graze if it is tied to a pole by a rope 7 m from its neck.
3. Calculate the circumference of a circle whose radius is $3\frac{1}{2}$ cm.
4. Find the circumference of a circle whose radius is 14 cm.
5. Calculate the area of a circle whose circumference is 154 cm.
6. Find the distance covered by a bike, which runs once round a circular track of radius 21 m.
7. Calculate to the nearest whole number the area of the shaded portion in the figure below.

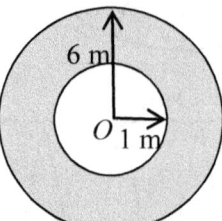

8. Two circles have radii 3.5 cm and 7 cm respectively. Find the area between them given that they have the same centre.

 Multiple Choice Exercise 11

1. The statement which is always true of a rhombus is:
   [A] All the angles are complementary   [B] All the sides are equal
   [C] The adjacent angles are equal   [D] All the angles are equal
2. The plane shape which is not a quadrilateral is:
   [A] Kite   [B] Rhombus   [C] triangle   [D] Parallelogram
3. The plane shape which is not an example of a quadrilateral is:
   [A] Square [B] Trapezium   [C] Rhombus   [D] Triangle
4. The statement which true of a quadrilateral is:

[A] The sum of its angles is 360°   [B] All its angles are equal.

[C] Two of its sides must be equal.   [D] It has more than 4 sides.

5. The largest angle of any triangle:

   [A] must always be an acute angle.   [B] can sometimes be an acute angle.

   [C] can never be a right-angle.   [D] must always be an obtuse angle.

6. A quadrilateral with one pair of sides parallel is:

   [A] a rhombus   [B] a parallelogram   [C] a rectangle   [D] a trapezium

7. The assertion about a rhombus which may not be true is:

   [A] The diagonals are equal.

   [B] The diagonals bisect the angles through which they pass.

   [C] The opposite angles are equal.

   [D] Opposite sides are equal.

8. By its angles and its sides the triangle in the following figure is:

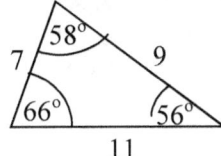

   [A] isosceles, obtuse   [C] isosceles, acute

   [B] scalene, obtuse   [D] scalene, acute

9. An irregular quadrilateral can never be:

   [A] a square   [B] a trapezium   [C] a rhombus   [D] a rectangle

10. The triangle in figure (a) below is:

    [A] isosceles   [B] equilateral   [C] scalene   [D] obtuse

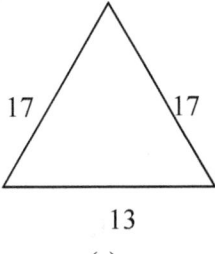

(a)

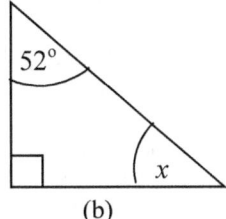

(b)

11. The value of x in the triangle in (b) above is:

    [A] 38°   [B] 142°   [C] 128°   [D] 232°

12. The sum of the angles of a square is:

    [A] 90°   [B] 120°   [C] 180°   [D] 360°

13. The triangle with angles 90°, 38°, and 52° is:

    [A] acute   [B] right   [C] obtuse   [D] reflex

14. Two angles of a triangle are 130° and 27°. The third angle and the type of triangle respectively are:

    [A] 203°, acute   [B] 23°, acute   [C] 23°, obtuse   [D] 203°, obtuse

15. A triangle with angles 49°, 86°, 45° is:

    [A] equilateral   [B] scalene   [C] isosceles   [D] obtuse

16. The shaded portion in the figure below is called:

[A] a minor segment   [B] a major segment
[C] a minor sector   [D] a major sector

17. A chord which bisects a circle is called:

[A] tangent   [B] segment   [C] diameter   [D] semi-circle

18. A quadrilateral whose diagonals bisect at right angles is:
[A] a rectangle   [B] a parallelogram   [C] a rhombus   [D] a square

19. The best name of the quadrilateral which has 4 congruent sides is:
[A] parallelogram   [B] rhombus   [C] trapezoid   [D] rectangle

20. The property/properties which do not characterize a rectangle is/are:

I. The diagonals bisect at right angles

II. Opposite sides are equal and parallel

III. Each of its angles is a right angle

[A] I only   [B] II only   [C] III only   [D] II and III only

21. The real name of the plane figure below and some of its possible names are:

[A] parallelogram; quadrilateral, rhombus.
[B] rhombus; quadrilateral, trapezium.
[C] trapezium; quadrilateral, polygon.
[D] quadrilateral, parallelogram, polygon.

22. A triangle has sides 3, 5, 8 and angles 25°, 85°, 70°. By the measure of its angles and its sides the triangle is:
[A] isosceles, obtuse   [B] scalene, obtuse   [C] isosceles, acute   [D] scalene, acute

23. The property which makes a rhombus different from every other parallelogram is:
[A] all sides are equal.   [B] Opposite sides are parallel.
[C] Opposite angles are equal.   [D] Diagonals bisect each other at right angles.

24. The property which makes a square a unique rectangle is:
[A] all sides are equal.   [B] Opposite sides are parallel.
[C] Opposite angles are equal.   [D] Diagonals bisect each other at right angles.

25. The property which makes a square a unique rhombus is:
[A] all sides are equal.   [B] Opposite sides are parallel.
[C] all angles are equal.   [D] Diagonals bisect each other at right angles.

26. The property which makes a rectangle a special parallelogram is:
[A] Opposite sides are equal.   [B] Opposite sides are parallel.
[C] Diagonals bisect each other.   [D] Diagonals are equal in length.

27. In the figure (a) below, the value of the angle marked *y* is:

[A] 28°   [B] 62°   [C] 118°   [D] 152°

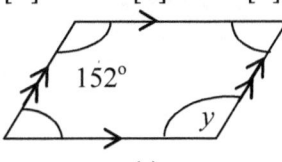
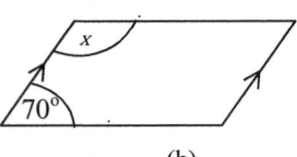

(a)   (b)

28. The value of the angle marked *x* in figure (b) above is:

[A] 140°   [B] 130°   [C] 110°   [D] 50°

29. The value of the angle marked *y* in the figure (a) below is:

[A] 140°   [B] 130°   [C] 110°   [D] 50°

30. In the figure (a) below, the sum of *x* and *y* is:

[A] 180°   [B] 210°   [C] 190°   [D] 270°

31. The values of *x*, *y*, and *z* in the figure (a) below are respectively:

[A] 130°, 50°, 130°   [B] 140°, 40°, 140°
[C] 150°, 30°, 150°   [D] 120°, 60°, 120°

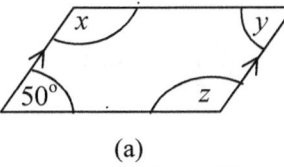
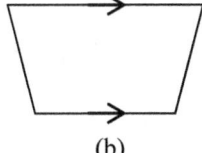

(a)   (b)

32. The name given to figure (b) above is:

[A] a parallelogram   [B] a trapezium   [C] a rhombus   [D] a rectangle

33. The regular quadrilateral among the following is:

[A] a square   [B] a rhombus   [C] a trapezium   [D] a rectangle

Topic 12

SYMMETRY

Objectives

At the end of this topic, the learner should be able to:

1. State the properties of line symmetry.
2. Locate the image of an object under line symmetry.
3. Identify figures with line or mirror symmetry.
4. Identify line symmetry in various polygons.
5. Identify polygons without line symmetry.
6. State the properties of point symmetry.
7. Locate the image of an object under point symmetry.
8. Identify figures with rotational and point symmetry.
9. Appreciate that point symmetry is rotational symmetry of even order.
10. Identify point symmetry and rotational symmetry in various polygons.
11. Identify polygons without point symmetry.

12.1 The Concept of Symmetry

Symmetry is the correspondence of parts on opposite sides of a point, line or plane.

Symmetry is a very important phenomenon in disciplines such as architecture, mathematics, biology, physics, mineralogy etc. The bodies of many animals for instance exhibit bilateral symmetry on two opposite sides of a linear axis, or a median plane.

Wat Phra Kaeo

The Wat Phra Kaeo temple built in 1782 is one among more than 18,000 Buddist temples in Thailand. It is one of the greatest symmetrical architectural designs in the world.

197

12.2 Line Symmetry

> **?** **Brainstorming Exercise**
>
> Consider the triangle *ABC* in the following figure.
>
>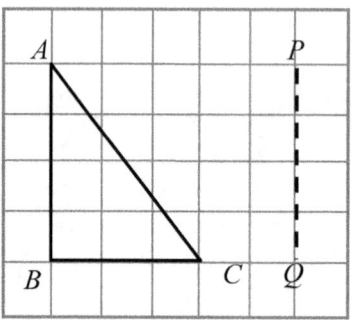
>
> Suppose a plane mirror is placed at *PQ*.
> 1. How will this triangle be seen in the mirror?
> 2. On a square paper, draw the triangle and its shadow in the mirror.

The following figure shows the triangle *ABC* called the **object** and its mirror ghost *A'B'C'* called the **image**. The line *PQ* on which a mirror is placed is called the **line of symmetry** and the phenomenon is called a **reflection, mirror symmetry,** line **symmetry** or **orthogonal symmetry.**

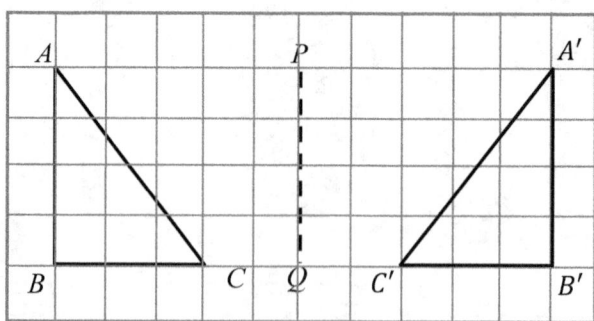

Certain plane figures are such that there is a line or many lines through which one half looks exactly like the other. For example a rectangle exhibits this property as shown in the figure below.

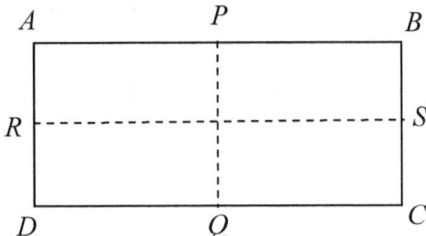

Properties of Line symmetry

 Investigative Activity

1. Draw each of the following plane figures on a typing sheet and show its lines of symmetry.
   (a) An isosceles trapezium (b) A kite   (c) An isosceles triangle
2. Identify all the corresponding points on either sides of the line of symmetry.
3. Measure the distance of each point and the corresponding point from the line of symmetry.
4. What conclusion do you draw?
5. Cut out your shape and fold it along the line of symmetry.
6. What do you notice concerning the size, shape and angles of one side and those of the other side of the line of symmetry?
7. Try to move the line of symmetry from its real position. Is the symmetry maintained?

From the above investigation, we conclude that:
(i) The line of symmetry remains unchanged. We say the line of symmetry is invariant.
(ii) The size of one halve is exactly the same as that of the other. In other words the image size and object size are the same.
(iii) Corresponding points on the two halves are the same distance from the line of symmetry. In other words corresponding points on the object and image are the same distance from the line of symmetry.
(iv) The shape of one halve is exactly the same as that of the other. In other words image shape and object shape are the same.
(v) Corresponding angles on the two halves or on the object and image are equal.

From the above line symmetry can therefore be defined as that property of a plane figure in which one of the halves of the object appears like the other part when viewed from a central line called the **line of symmetry**.

In the four-pointed star and regular octagon in the following figure, the line of symmetry is shown by the dotted lines. The star has four lines of symmetry. Can you count them? The regular octagon has eight lines of symmetry. Also, count them.

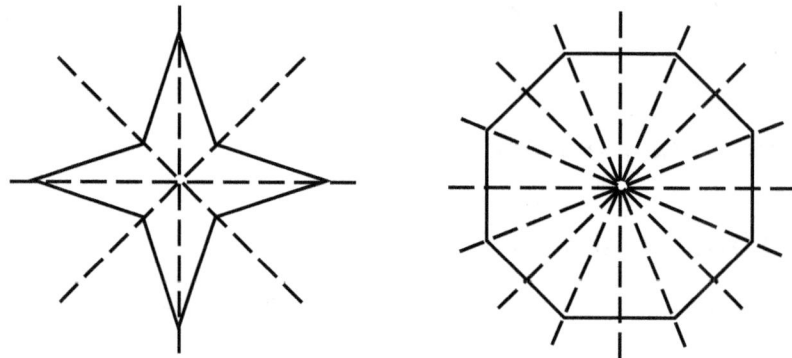

The following table shows more plane figures and their lines of symmetry. The lines of symmetry are shown dotted.

Plane Figure	Diagram	Number of lines of symmetry
Isosceles trapezium		1
Isosceles triangle		1
Kite		1

Plane Figure	Diagram	Number of lines of symmetry
Rhombus		2
Rectangle		2
Equilateral triangle		3
Square		4
Parallelogram		none

 ## Exercise 12:1

1. By drawing and showing using dotted lines state the number of lines of symmetry, if any which the following have.
   (a) An isosceles trapezium   (b) A kite   (c) An isosceles triangle
   (d) A rhombus   (e) A rectangle   (f) An equilateral triangle
   (g) A parallelogram   (h) A square   (i) A circle
2. Name all the capital letters of the English alphabet which have one and only one line of symmetry
3. List all the capital letters of the English alphabet, which have at least two lines

of symmetry.
4. List all the capital letters of the English alphabet, which have no lines of symmetry.
5. Identify and draw all the lines of symmetry in the following plane shapes.

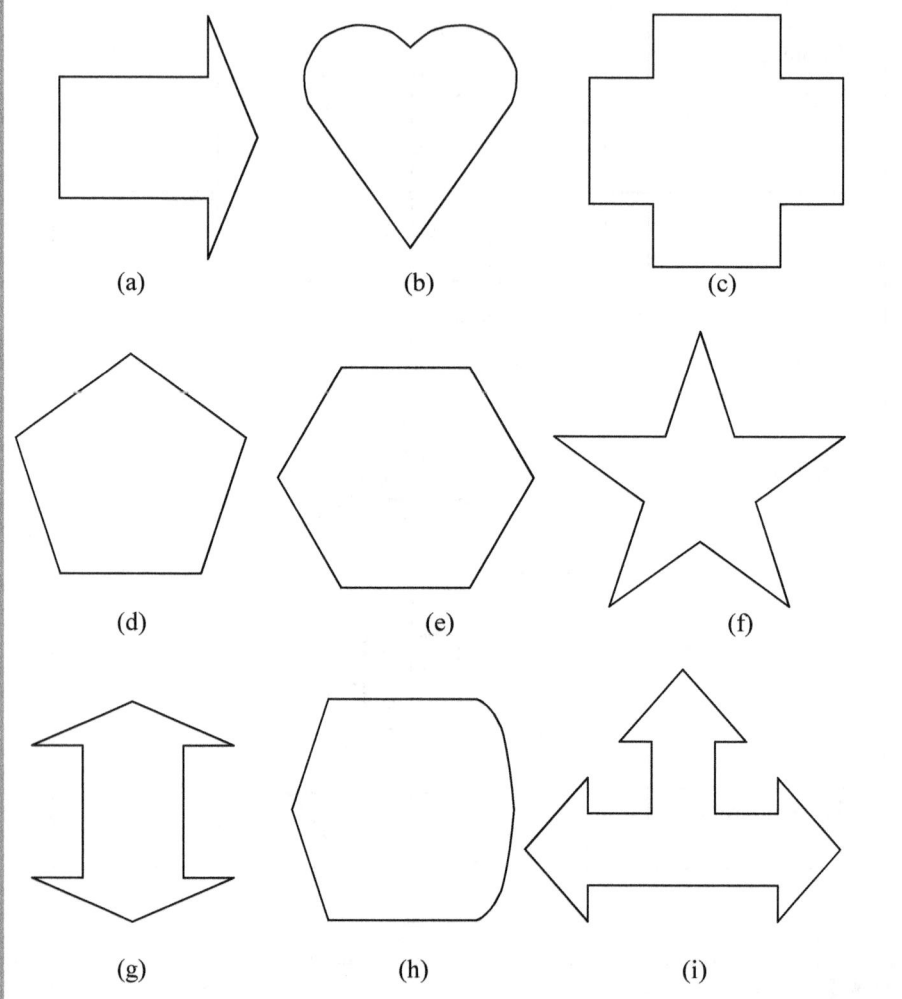

| (a) | (b) | (c) |

| (d) | (e) | (f) |

| (g) | (h) | (i) |

6. Make a collection of five different flat objects from your environment which exhibit line symmetry. You will be required to show the lines of symmetry of your collected objects to your teacher and class.

12.3 Point Symmetry or Radial Symmetry

> **?** | **Brainstorming Exercise**
>
> Consider the following German army symbol called swastika.
>
>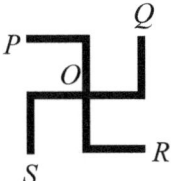
>
> 1. Does the figure have any line of symmetry?
> 2. Suppose that the symbol is rotated through 90° clockwise or anticlockwise about the point of intersection *O*, will there seem to be any recognizable change in the shape, size, position and angles on the figure?
> 3. In how many positions can the symbol be rotated so that there won't appear to be any recognizable change in shape, size, position and angles?

Clearly the symbol swastika has no line of symmetry. However it exhibits some properties which are worth noting. If the symbol is rotated through 90° clockwise or anticlockwise about the point of intersection *O*, there seem to be no recognizable change in the shape, size, position and angles on the figure. There are four positions through which we can rotate the symbol so that there won't appear to be any recognizable change in shape, size, position and angles.

A plane figure which exhibits the property that when rotated about a point in its plane through an angle does not seem to have any recognizable change in shape, size, position and angles is said to have **point** or **rotational or radial symmetry**. In the case of swastika there are four positions through which we can rotate the figure, so that the figure does not seem to have any recognizable change in shape, size, position and angles. In this case we say that swastika exhibits rotational symmetry of **order** four.

Examine the figures (a), (b) and (c) below.

The five point star in figure (a) has rotational symmetry of order five.

A circle (figure (b)) rotated any amount about the center, remains unchanged. Therefore, a circle exhibits rotational symmetry of infinite order about its center.

An equilateral triangle (figure (c)) remains unchanged in three different positions when rotated about the point of symmetry. Hence, an equilateral triangle exhibits point symmetry of order 3.

Notice that these plane figures equally exhibit line symmetry. Therefore some plane figures can exhibit both line and rotational symmetry.

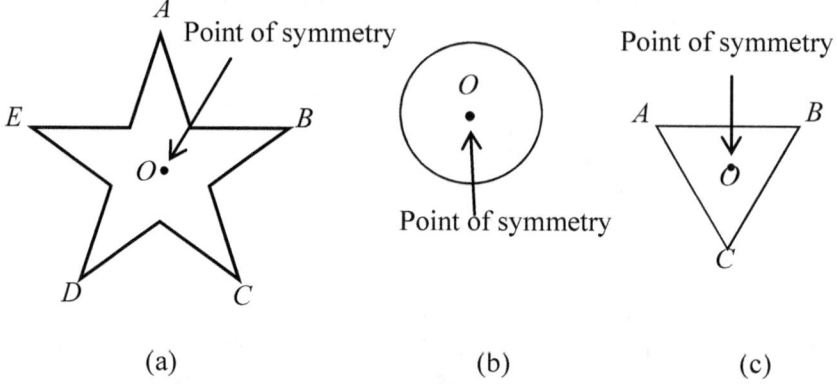

| (a) | (b) | (c) |

Any regular plane figure (figure with equal sides and angles) has rotational symmetry of order equal to its number of sides as shown in the following table.

Regular Plane Figure	Diagram of shape	Order
Equilateral triangle		3
Square		4
5 sided plane figure		5
6 sided plane figure		6

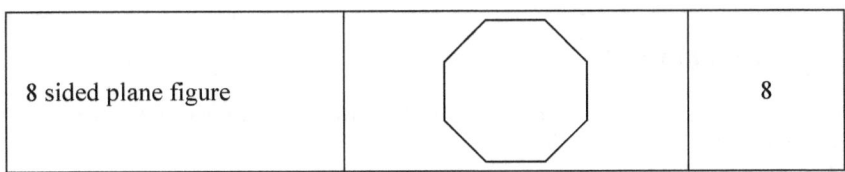

| 8 sided plane figure | | 8 |

Rotational symmetry of even order is called **point symmetry**. The image of a figure under point symmetry about centre O is a figure that can be superimposed on the original by rotating it a half-turn, about O.

The following figure shows that a parallelogram exhibits point symmetry about the point O. A_2, B_2, C_2, and D_2 can respectively fit on A_1, B_1, C_1 and D_1 if turn through a half turn or 180°. In order words if the parallelogram is rotated through 180°, it appears to be unchanged.

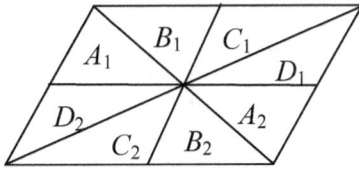

Properties of Point or Radial Symmetry

If we conduct the above investigation for rotational symmetry, we see that all the properties of line symmetry are true for rotational symmetry. Thus,

(i)   The line of symmetry remains unchanged. We say the line of symmetry is invariant.

(ii)   The size of one halve is exactly the same as that of the other. In other words the image size and object size are the same.

(iii)   Corresponding points on the two halves are the same distance from the line of symmetry. In other words corresponding points on the object and image are the same distance from the line of symmetry.

(iv)   The shape of one halve is exactly the same as that of the other. In other words image shape and object shape are the same.

(v)   Corresponding angles on the two halves or on the object and image are equal.

Exercise 12:2

1. By drawing and showing using dots state the order of rotational symmetry, if any which the following exhibit.
   - (a) An isosceles trapezium
   - (b) A kite
   - (c) An isosceles triangle
   - (d) A rhombus
   - (e) A rectangle
   - (f) An equilateral triangle
   - (g) A parallelogram
   - (h) A square
   - (i) A circle
2. List all the capital letters of the English alphabet, which exhibit rotational symmetry of order at least two.
3. List all the capital letters of the English alphabet, which exhibit no rotational symmetry.
4. Identify and draw diagrams showing the point of symmetry if any in the following plane shapes. In each case state the order of rotational symmetry.

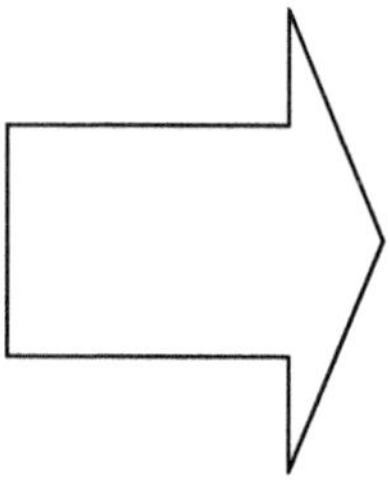

(a)

(b)

(c)

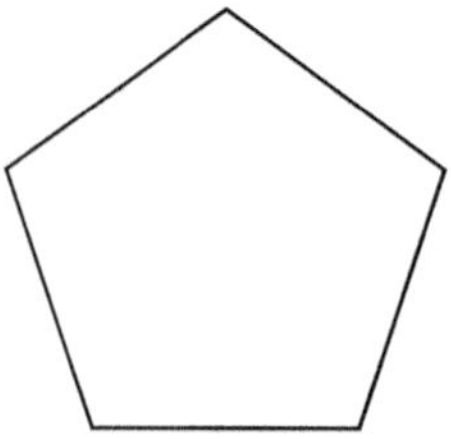

(d)

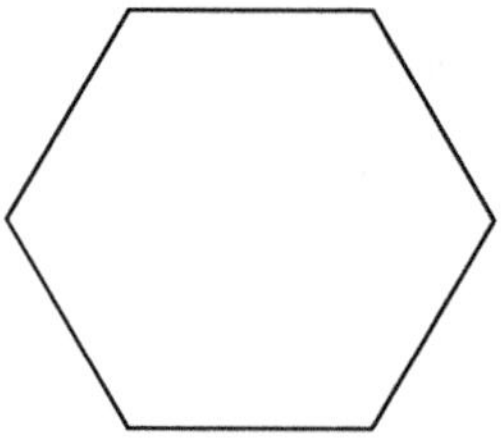

(e)

(f)

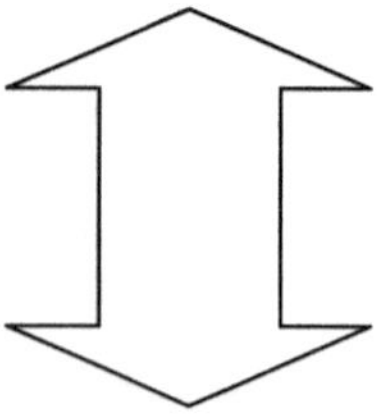

(g)

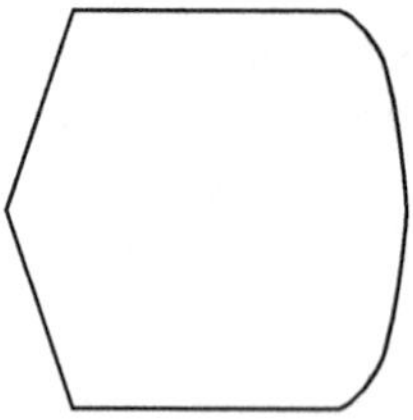

(h)

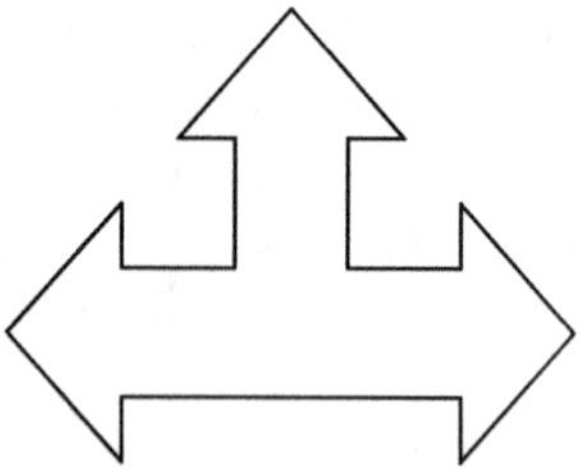

(i)

5. Make a collection of five different flat objects from your environment which exhibit rotational symmetry. You will be required to show the point of symmetry of your collected objects to your teacher and class.

6. Complete each of the diagrams in the following figure, so that each will have rotational symmetry of order (i) two (ii) three (iii) four

(a)   (b)   (c)

 Integration Activity

Make three different designs of a ceiling, a floor or a dress using as many combinations of plane shapes as you desire. Each design should portray both line and rotational symmetry.

 Multiple Choice Exercise 12

1. Using their symmetric properties, the odd plane figures is:
   [A] An isosceles triangle.   [B] a semi-circle.
   [C] a rectangle.   [D] a pentagon with four sides equal.
2. The number of lines of symmetry in a rectangle is:
   [A] 1   [B] 2   [C] 4   [D] 8
3. Symmetrically a square differs from a rectangle because:
   [A] a square has 2 lines of symmetry while a rectangle has 4.
   [B] a square has 4 lines of symmetry while a rectangle has 2.
   [C] the 4 sides of a square are equal but a rectangle has a pair of opposite sides.
   [D] the diagonals of a square intersect at right angles but those of a rectangle do not.
4. The number of letters in the word MATHEMATICS which possess point symmetry is:
   [A] 0   [B] 1   [C] 2   [D] 3
5. The letter which possesses point symmetry is:
   [A] M   [B] Z   [C] Y   [D] E
6. The quadrilateral which has exactly one line of symmetry is:
   [A] a kite   [B] a rectangle   [C] a parallelogram   [D] a rhombus
7. The graph which shows the reflection of triangle *ABC* in the line *XY* in figure (a) below is:

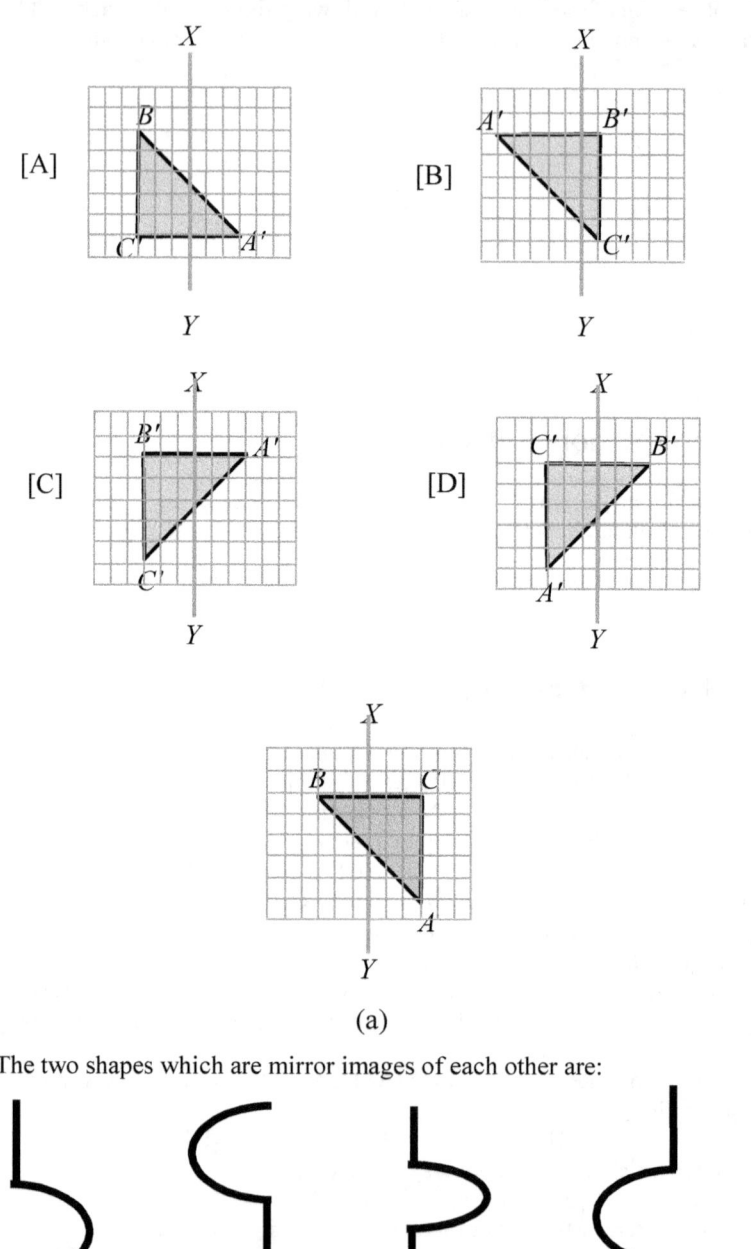

[A]

[B]

[C]

[D]

(a)

8. The two shapes which are mirror images of each other are:

I.   II.   III.   IV.

[A]   I and II   [B] I and III   [C]   I and IV   [D] II and IV

9. The figure that possesses rotational symmetry is:

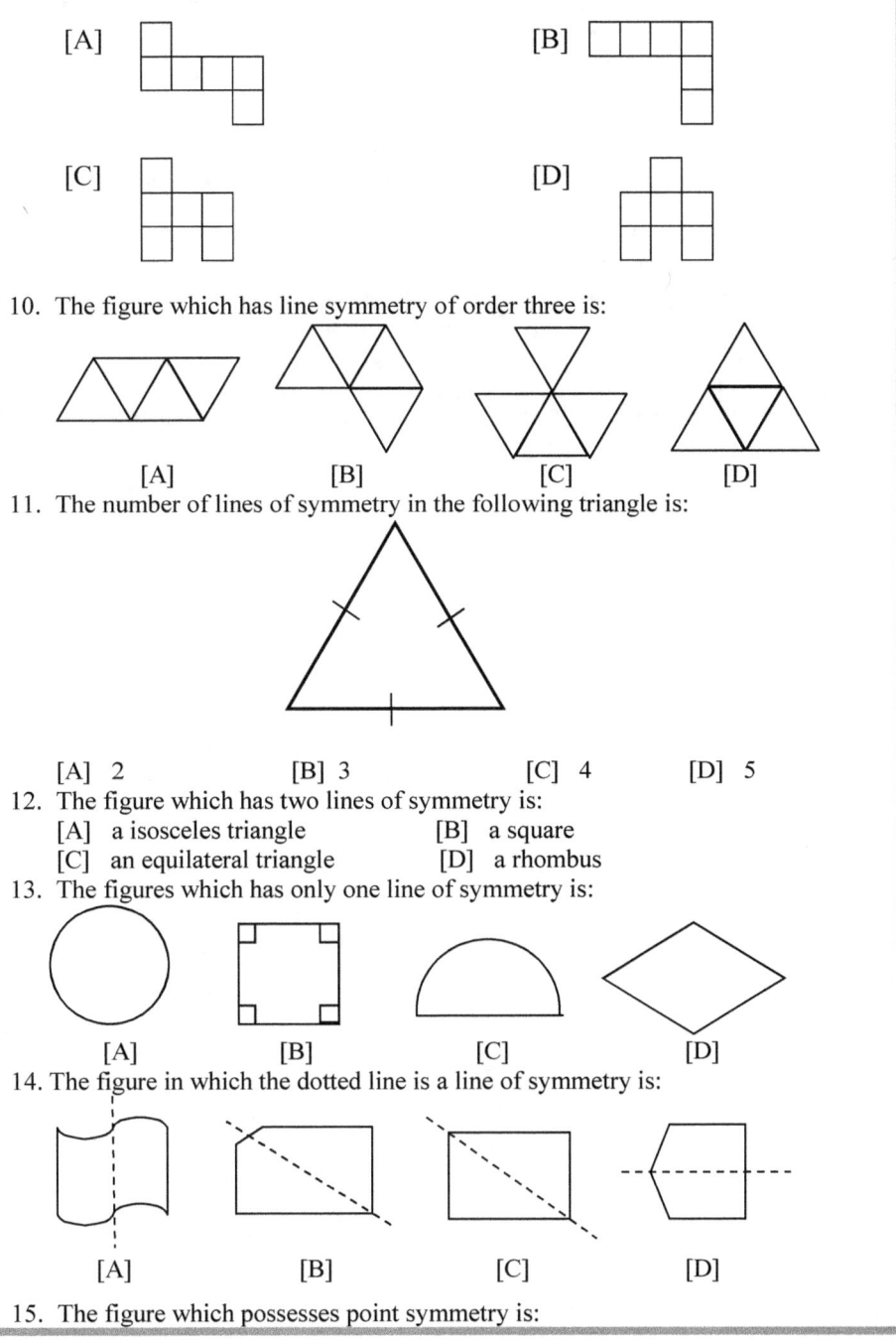

[A]   [B]

[C]   [D]

10. The figure which has line symmetry of order three is:

　　[A]　　　[B]　　　[C]　　　[D]

11. The number of lines of symmetry in the following triangle is:

　　[A] 2　　　[B] 3　　　[C] 4　　　[D] 5

12. The figure which has two lines of symmetry is:
　　[A]　a isosceles triangle　　　[B]　a square
　　[C]　an equilateral triangle　　[D]　a rhombus

13. The figures which has only one line of symmetry is:

　　[A]　　　[B]　　　[C]　　　[D]

14. The figure in which the dotted line is a line of symmetry is:

　　[A]　　　[B]　　　[C]　　　[D]

15. The figure which possesses point symmetry is:

209

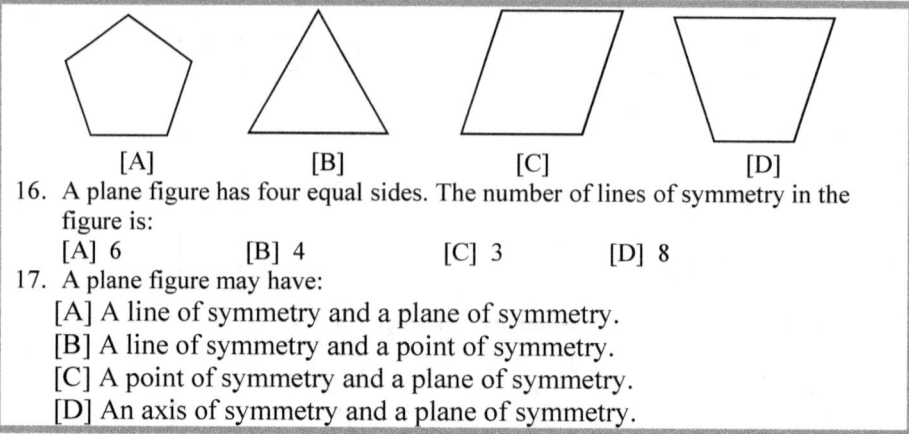

| [A] | [B] | [C] | [D] |

16. A plane figure has four equal sides. The number of lines of symmetry in the figure is:

[A] 6   [B] 4   [C] 3   [D] 8

17. A plane figure may have:

[A] A line of symmetry and a plane of symmetry.

[B] A line of symmetry and a point of symmetry.

[C] A point of symmetry and a plane of symmetry.

[D] An axis of symmetry and a plane of symmetry.

Topic 13

COORDINATE GEOMETRY

Objectives

At the end of this topic, the learner should be able to:

1. Draw and label the Cartesian plane.
2. Identify and distinguish between the abscissa and the ordinate.
3. Choose and use a scale to graduate and calibrate the axes.
4. Plot and read points on the Cartesian plane.
5. Identify points which are on the same horizontal line or which are on the same vertical line.
6. Choose and graduate a scale.
7. Apply coordinates to real life situations.

13.1 The Concept of Coordinates

The following figure is the plan of a classroom. Each student has a separate table. The tables are arranged into columns and rows. The columns and rows are numbered, so that it is possible to identify a student by his column number and row number.

4 Feh	Tah	Abe	Ndeh	Nkeh	
3 Ndi	Nfor	Kofi	Mah	Jiti	
2 Neh	Lum	Nfih	Tata	Ndah	Nfu
1 Jum	Seh	Nuh	Abu	Komi	Bani
1	2	3	4	5	6

Brainstorming Exercise

Identify the student who sits on
(a) Column 2, row 3.   (b) Column 6, row 2.
(c) Column 4, row 1.   (d) Column 3, row 4.

Instead of writing column 3, row 2; (3,2) in that order can be written, to stand for column 3, row 2, making it a rule to write the column number first and row number second in the pair of numbers. Following this rule, (3,4) and (4,3) do not mean the same. The order in which the numbers are written is therefore very important. A pair of numbers written in this way is called an **ordered pair**, because they are in pairs and their order is very important. A set of numbers or a single number that locates a point on a line, on a plane, or in space such as (3,2) is called a **coordinate**.

Exercise 13:1

1. The ordered pairs stand for column number and row number in that order. With reference to the figure above, write down the meaning of the following.
   (a) (4,3)   (b) (4,1)   (c) (2,2) (d) (3,4)   (e) (1,4)   (f) (4,2)
2. Say whether or not (6,1) and (1,6) mean the same giving reasons for your answers.

? | Brainstorming Exercise

1. What are the advantages of using a plane such as the one in the figure above?
2. Name at least ten other situations in which we can use a plan similar to the one in the figure above.

13.2 Real Life Applications of Coordinates

The idea of coordinates is very useful in real life. In plantations, crops are usually planted in rows and columns. Surveyors use it to plan towns and architects use it to plan houses. Cartographers and draftsmen use it for designing different plans to scale. Librarians and store keepers use the principle to arrange books and other articles on shelves. Games such as scrabble, monopoly, draught and chess are designed based on this principle. Dress makers employ the principle to design dresses. Coaches in some field games such as football use the principle to draw line ups. The idea is used in locating places on the earth's surface by employing longitudes and latitudes. In fact the list of situations in which coordinates are used is inexhaustible.

13.3 Plotting of Points

We can simplify the plan in the figure above as shown in the figure, *C* standing for column and *R* for row. To represent the position of a student, a cross ×, is marked at the point where his column number intersects his row number. The position of Abe has been done as an example.

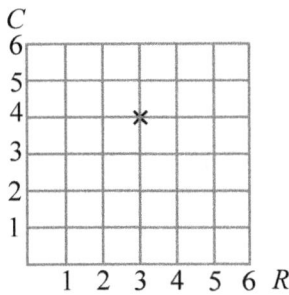

Map of West Cameroon

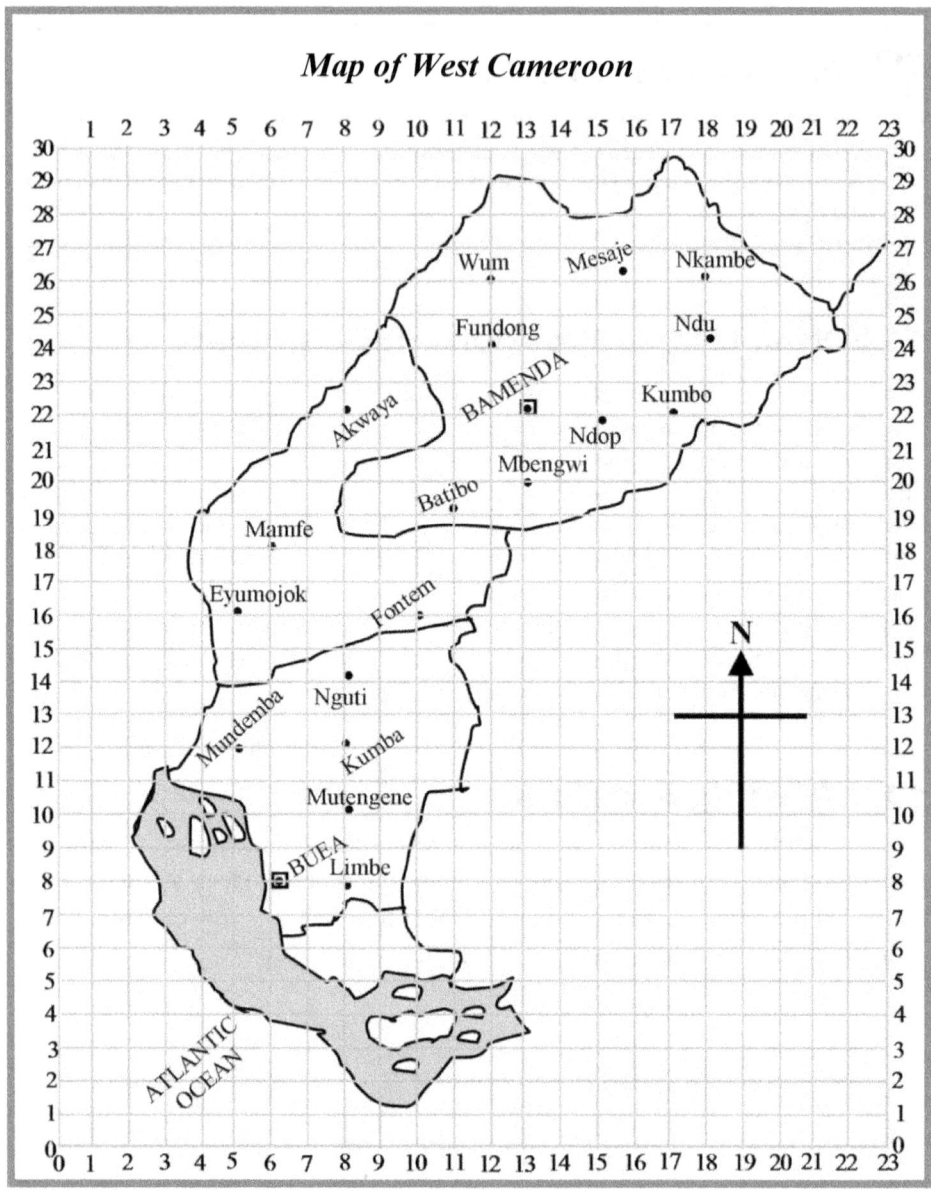

Exercise 13:2

1. In the figure below, write down the ordered pairs representing the four unlabeled points and label against each the names of the students who sit there.

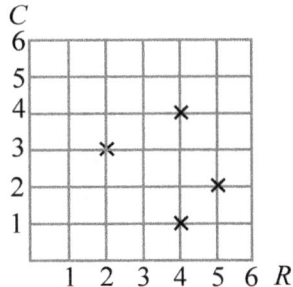

2. Draw another diagram and mark the points where Jum, Tata, Feh, Bani sit, using the first letters of their names.
3. The figure below is a map of West Cameroon. Using the grid lines on the Map, write down the ordered pairs representing the following towns; Bamenda, Limbe, Mbengwi, Mutengene, Kumba, Fundong, Nkambe, Buea, Nguti, Eyumojock. The first number is that of the vertical grid line while the second is that of the horizontal grid line. For instance the ordered pair representing Akwaya is (8,22).
4. Using the map below, state the town represented by the following ordered pairs. (12,26), (15,22), (10,16), (6,8), (17,22).

13.4 The Cartesian (or Coordinate) Plane

A **Cartesian plane** or **coordinate plane** or the *x-y* **plane**, is a diagram on which geometric figures are represented in two dimensions using a coordinate system. It was invented by the French Mathematician René Descartes (1596-1650). Diagrams like those which we used earlier to represent the sitting plan are examples of Cartesian planes. Generally the vertical bold line labeled *C* is labeled *y*, and called the *y*-**axis** while the horizontal bold line labeled *R* is labeled *x*, and called the *x*-**axis**. The *x*- and the *y*- axes meet at a point *O*, called the **origin**.

In this particular case, the first number in each ordered pair called the *x*-**coordinate** or **abscissa** is always taken from the *x*-axis, while the second called the *y*-**coordinate** or **ordinate** is taken from the *y*-axis. The ordered pairs are called **coordinates**. Notice that these axes are **perpendicular** to each other (i.e. they meet at 90^0 to each other). Another name for perpendicular is **orthogonal**.

Note that the numbers written against the x-axis refer to the vertical lines, while those written against the y-axis refer to the horizontal lines. Don't be confused with these lines!

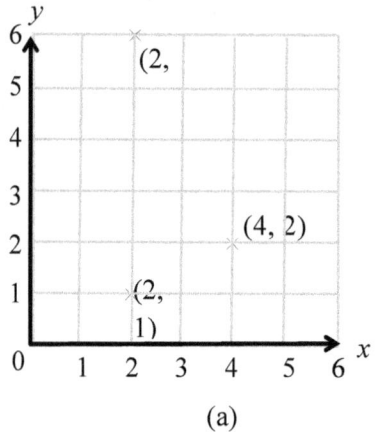

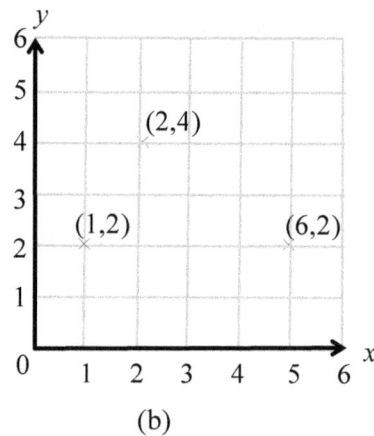

(a)   (b)

The figure above shows the three points (2,1), (4,2) and (2,6) plotted and labeled. It should again be emphasized that the points (2,1),(4,2) and (2,6) are different from the points (1,2), (2,4) and (6,2).

13.5 Extension of the Axes

We can visualize the x- and y- axes as vertical and horizontal number lines drawn to intersect at their origins and we can extend each of these axes in both directions as long as possible as shown in the following figure.

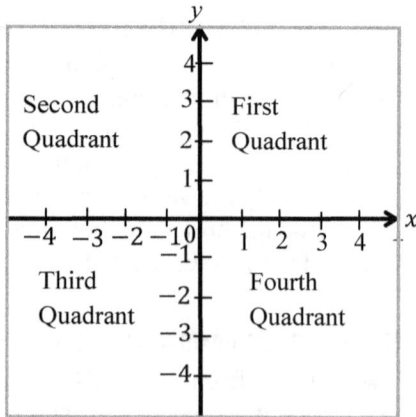

The x-axis and the y-axis divide the x-y plane into four sections called the first, second, third and fourth quadrants, as shown in the figure above. In the first quadrant, both the *x*- and *y*- coordinates are positive, in the second quadrant the *x*-coordinate is negative and the *y*- coordinate is positive in the third quadrant both are negative and in the fourth quadrant the *x*- coordinate is positive while the *y*- coordinate is negative.

The Cartesian plane below shows the points $(2,3)$, $(-3,0)$, $(-4,-1)$, and $(3,-2)$.

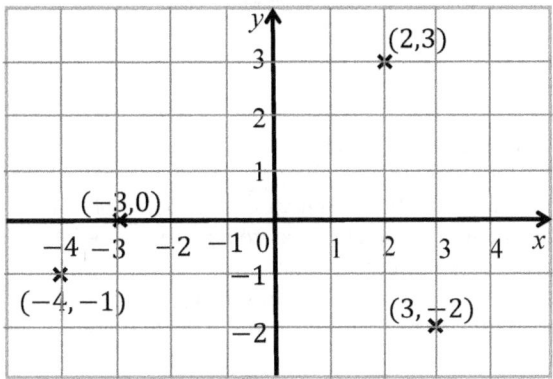

 Exercise 13:3

1. In which quadrant is each of the following points? $A(1,2)$, $B(1,-2)$, $C(-1,2)$, $D(-1,-2)$, $E(3,4)$, $F(4,-1)$, $G(-1,4)$ and $H(-4,-2)$.
2. Write down the coordinates of the points lettered in the following figure.

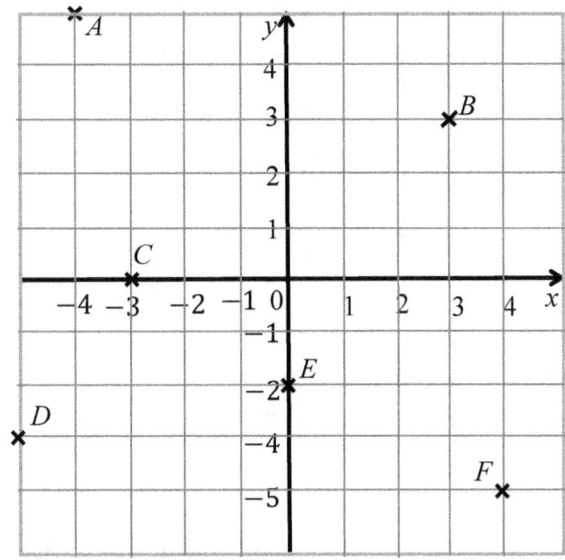

3. Plot the points $A(1,4)$, $B(4,1)$, $C(3,0)$ and $D(0,3)$ on the same Cartesian plane.

4. Plot on the Cartesian plane the points $W(0,-3)$, $X(-3,0)$, $Y(1,-2)$, $Z(-2,1)$.
5. Plot the points $P(-1,5)$, $Q(6,-5)$, $R(-3,-4)$ and connect them with straight lines.

13.6 Points on horizontal and Vertical Straight Lines

 Investigative Activity

1. In the following figure, state the coordinates of the points A, B and C.
2. What can you say about the ordinate (y-coordinate) of the points A, B and C?
3. State the coordinates of the points P, Q and R.
4. What can you say about the abscissa (x-coordinate) of the points P, Q and R?
5. Write down the coordinates of four points which lie on the same horizontal straight line.
6. Write down the coordinates of four points which lie on the same vertical straight line.

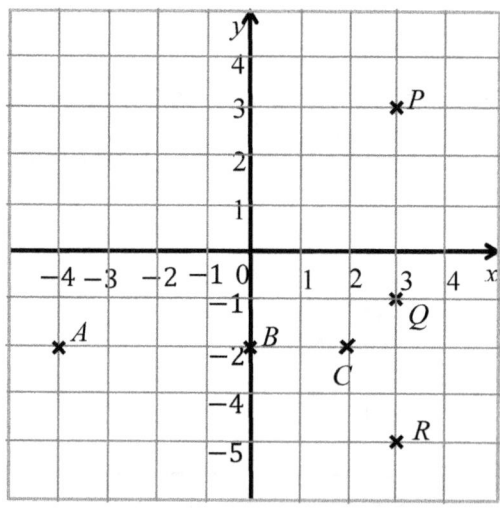

From the above figure, the coordinates of the points A, B and C are $(-4,-2)$, $(0,-2)$ and $(2,-2)$ respectively. The ordinate (or y-coordinate) of each of the points is -2.

Generally,

The ordinate of points which lie on the same horizontal straight line is the same.

218

Also the coordinates of the points P, Q and R. are (3,3), (3, −1) and (3, −5) respectively. The abscissa of each of the points is 3.

Generally,

The abscissa of points which lie on the same vertical straight line is the same.

 Exercise 13:4

1. State whether the straight line containing the given pair of points is horizontal, vertical or neither.
   (a) (5,2) and (2,5) (b) (−6,4) and (−6,13) (c) (7,10) and (−1,10)
   (d) (3,−8) and (−3,8) (e) (0,5) and (5,0) (f) (0,−12) and (0,14)
2. Three vertices of a rectangular hall are (−10,10), (−10,13) and (−7,10). State the coordinate of the fourth vertex.

13.7 Scaling

In cartography, scale is the ratio of the distance between two points on a map or model and the actual distance between the two points on the earth's surface or object. On a number line, Cartesian plane or graph a scale is the number of units, which represent a unit length (quantity) along the line or a definite axis.

? **Brainstorming Exercise**

1. Name five situations in which scales are used.
2. State the advantages and disadvantages of a small scale over a large scale.

13.8 Scale Representation

On maps, scale is represented in three ways:

1. As a ratio or fraction, such as 1:100,000 or $\frac{1}{100,000}$, which means that 1 unit of measurement on the map equals 100,000 of the same units on the earth's surface.
2. As a phrase in words and figures, such as "1 cm represents 100 km" (that is, 1 cm on the map represents 100 km on the earth's surface).
3. As a graphic scale, usually a straight line on which distances (most often in kilometers or miles) have been marked off as shown below.

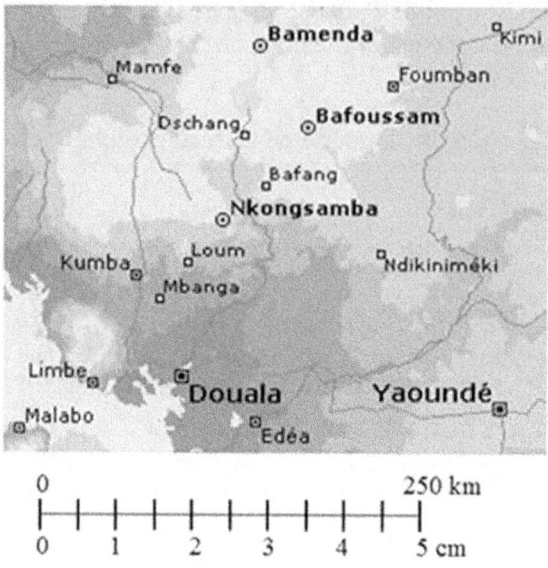

The larger the scale of a map, the closer it approaches the actual size of features on the earth's surface or on the actual object. Small scales generally show larger portions of the real object and have less detail than large scales.

Maps, scale drawings, models, plans of buildings, patterns and graphs are drawn to scale.

13.9 Graduating the Scale on the Cartesian Axes

On the Cartesian plane, if both the scale and the range are given, use them to graduate the axis (or axes) uniformly, (i.e. making sure that the scale for each axis is consistent all through the axis). Most good scales are often in powers of 10 or based on factors of powers of 10. For instance,

> 1 centimetre represents 1 unit,
>
> 1 centimetre represents 100 units,
>
> 1 centimetre represents 50 units,
>
> 1 centimetre represents 20 units,
>
> 2 centimetres represents 5 units, etc.

The scales on both the x- and y- axes may be the same or may be different.

Example

1. Taking 1 squares to represent 1 unit on both axes, plot the points $A(3,2)$, $B(-2,3)$, $C(1,-2)$, $D(-3,-1)$, $E(1,0)$

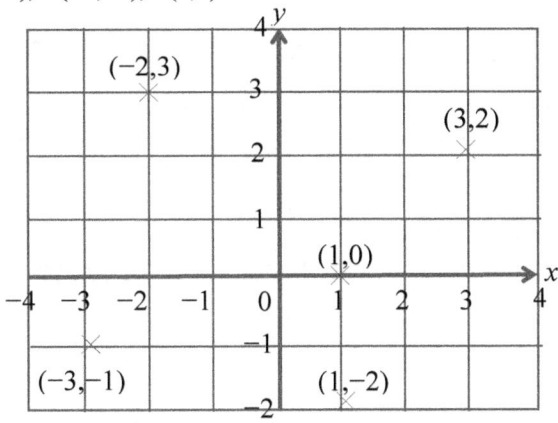

2. Using a scale of 2 squares to represent 1 unit on the x-axis and 2 squares to represent 2 units on the y-axis plot the points $A(-2,4)$, $B(-1,0)$ and $C(-3,-4)$ and link them up with straight lines. On the same Cartesian plane plot the points $X(2,4)$, $Z(1,0)$ and $Y(3,4)$ and link them up with straight lines.

Solution

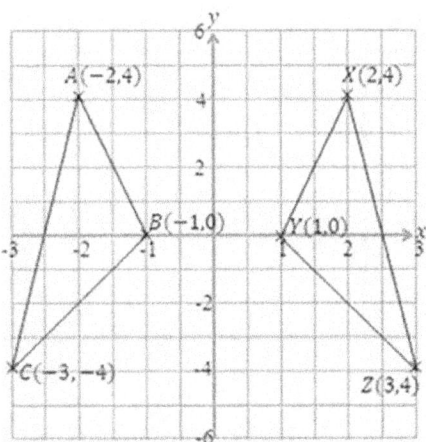

Graphs are plotted better using graph paper than square paper. The common types of graph papers are the 1 mm graph paper and the 2 mm graph paper. On the 1mm graph paper (figure (i) below), the thicker lines are half a cm apart.

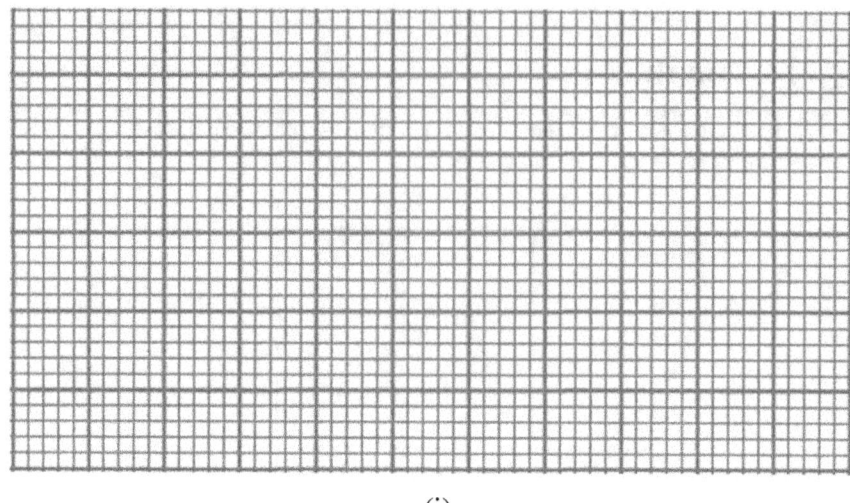

(i)

On the 2 mm graph paper (figure (ii) below), the thinnest lines are 2 mm apart. The thicker lines are a cm apart and the thickest lines are 2 cm apart.

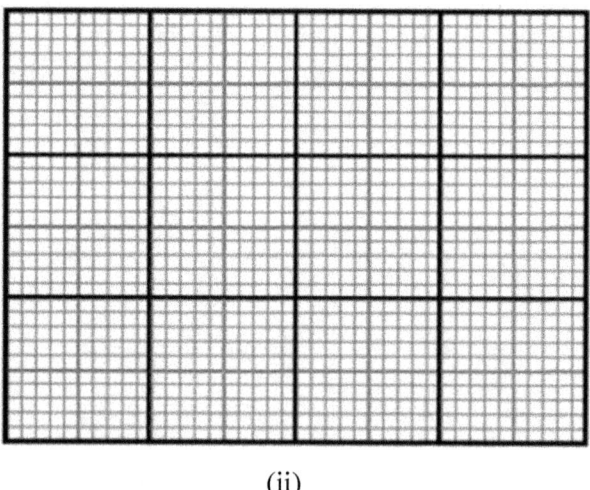

(ii)

Examples

1. Using a scale of 1 cm to represent 2 units on both axes plot the points $W(-2,6)$, $X(-6,-6)$, $Y(6,-6)$ and $Z(4,6)$ and link them with straight-line segments. What shape do you obtain?

Solution

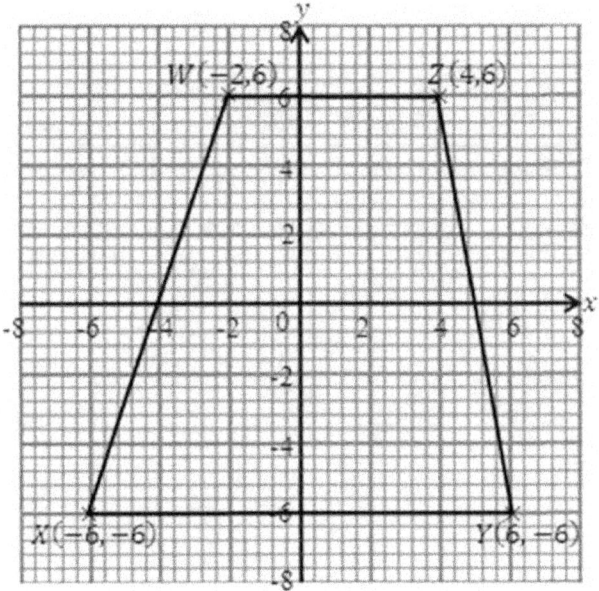

2. Using a scale of 1 cm to represent 2 units on the x-axis, and 1 cm to represent 4 units on the y-axis plot the points $A(-2,12)$, $B(-6,-4)$ and $C(-2,-8)$ and link them with straight-line segments. Also plot the points $P(4,8)$, $Q(2,0)$, $R(4,-12)$ and $S(6,0)$ on the same Cartesian axes and link them with straight-line segments. What shapes do you obtain?

Solution

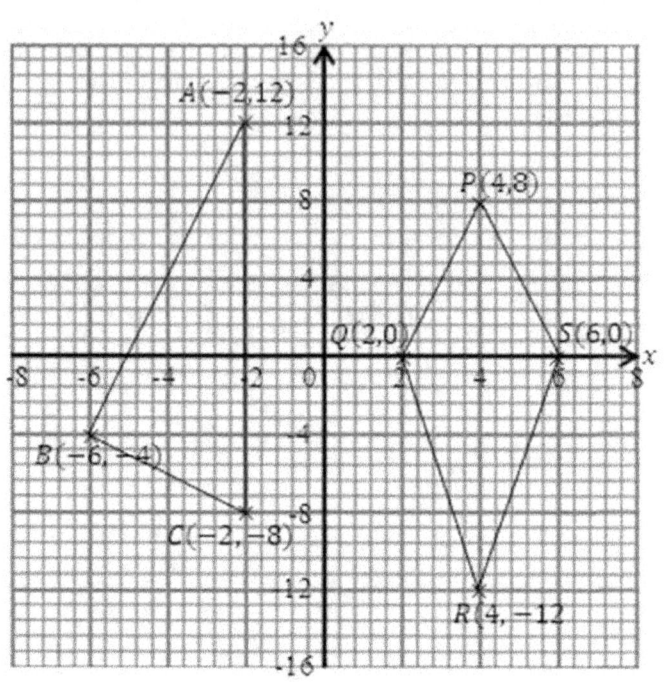

 Exercise 13:5

(1) Using 2 squares to represent 1 unit on both axes, plot the points $A(1,4)$, $B(4,1)$, $C(3,0)$ and $D(0,3)$ on the same Cartesian plane.

(2) Using 2 cm to represent 1 unit on the y-axis and 4 cm to represent 1 unit on the x-axis, plot the points $W(0,-3)$, $X(-3,0)$, $Y(1,-2)$ and $Z(-2,1)$ on the same Cartesian plane.

(3) Taking 2 cm to represent 1 unit on the x-axis and 1 cm to represent 2 units on the y-axis, plot on the Cartesian plane the points

(4) Using 2 cm to represent 1 unit on the y-axis and 4 cm to 1 unit on the x-axis, plot the points $P(-1,-1)$, $Q(-2,-3)$, $R(-4,-1)$ and connect them with straight lines.

13.10   Longitudes and Latitudes

In the map of west Cameroon an arbitrary grid was used to locate places on the map. The map of Africa below shows how geographers exploit the idea of coordinates to locate places on the earth surface using longitudes and latitudes. Longitudes are imaginary lines on the earth surface which run from north to south, east or west of the Greenwich meridian. Latitudes on the other hand, are imaginary lines on the earth surface which run from east to west, north or south of the equator.

MAP OF AFRICA SHOWING LONGITUDES AND LATITUDES

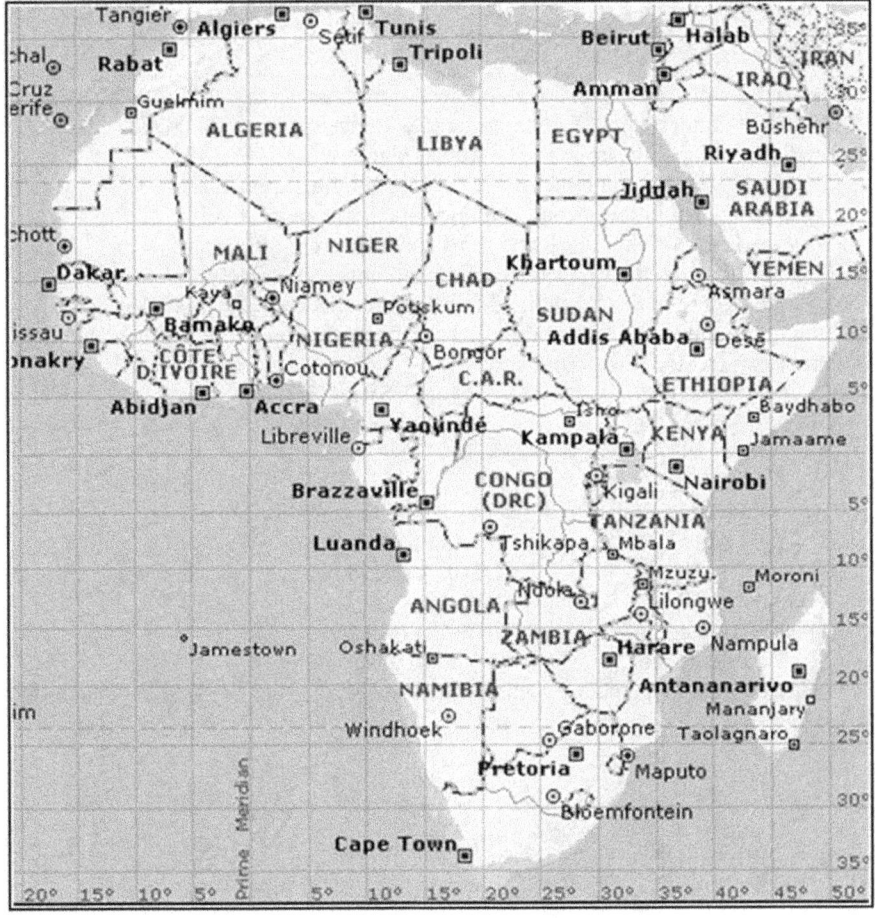

Exercise 13:6

1. Given that the first number in the ordered pair stands for the longitude and the second for the latitude. For example $(-3, -15)$ means longitude 3° W and 15° S. Write down the town which is located on
   (a) $(39, -15)$ (b) $(15,10)$ (c) $(15, -5)$ (d) $(18,15)$
   (e) $(32,0)$ (f) $(18, -34)$ (g) $(-15,10)$ (h) $(10,37)$
2. State the longitude and latitude of the following as ordered pairs.
   (a) Gaborone (b) Accra (c) Yaoundé (d) Brazzaville

 Multiple Choice Exercise 13

1. A triangle with vertices at the points with coordinates $(-4,4)$, $(4,4)$, $(1,-1)$ is:
   [A] Right–angled [B] Equilateral [C] Isosceles [D] Scalene
2. The following points are in the Cartesian plane. $P(-1, -4)$, $Q(6, -5)$, $R(-1,5)$, $S(3,2)$. The statement which is true about the points is:
   [A] R and Q are in the second and third quadrants respectively.
   [B] P and S are in the fourth and third quadrants respectively.
   [C] S and R are in the first and second quadrants respectively.
   [D] P and Q are in the second and fourth quadrants respectively.
3. Among the following, the straight line which is parallel to the x-axis is the line that passes through the points:
   [A] $(-1,5)$ and $(2,0)$ [B] $(3,-5)$ and $(3,3)$
   [C] $(-2,1)$ and $(1,1)$ [D] $(0,0)$ and $(3,6)$
4. The point which lies on the y –axis is:
   [A] $(0,-1)$ [B] $(2,-5)$ [C] $(1,0)$ [D] $(3,1)$
5. The straight line parallel to the y-axis passes through the points:
   [A] $(-1,5)$ and $(2,0)$ [B] $(3,-5)$ and $(3,3)$
   [C] $(-2,1)$ and $(1,1)$ [D] $(0,0)$ and $(3,6)$
6. Given that $x > 0$ and $y < 0$ the point (x, y) lies:
   [A] second quadrant   [B] first quadrant
   [C] fourth quadrant   [D] third quadrant
7. On the coordinate plane the points $A(1, -3)$, $B(-1,4)$, and $C(-2, 3)$ can be plotted as:

[A]
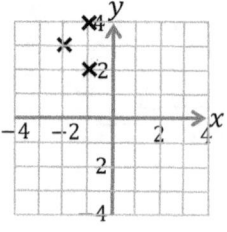

[B]
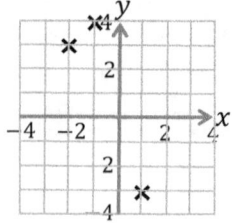

[C] 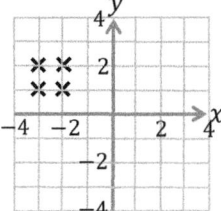   [D]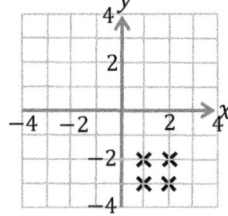

8. *ABCD* has vertices *A*(–3, 1), *B*(–3, 2), *C*(–2, 1) and *D*(–2, 2). The graph of *ABCD* is:

[A] 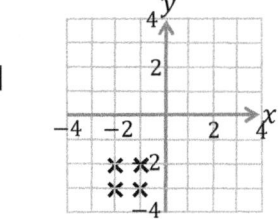   [B]

[C]   [D]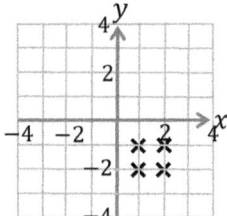

9. A rectangular playground on a scale drawing has vertices (–2, 1), (–2, 5) and (0, 1). The coordinates of the fourth vertex are:
    [A] (5, 0)   [B] (0, 5)   [C] (–4, 5)   [D] (0, –3)

10. The coordinates of point *A* in the figure below are:
    [A] (–2, 3)   [B] (3, –2)   [C] (–2, –3)   [D] (2, 3)

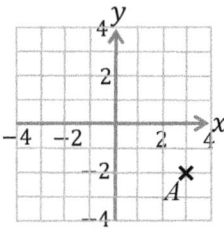

11. The graph of the line containing the points (2, –3) and (0, –3) is:
    [A] Horizontal like this   [B] Vertical like this

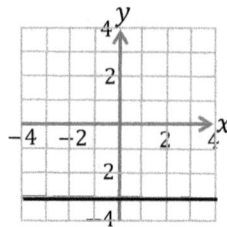

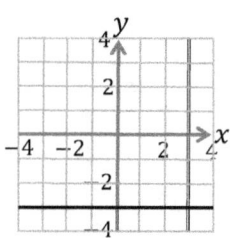

[C]   Vertical like this   [D]   Horizontal like this

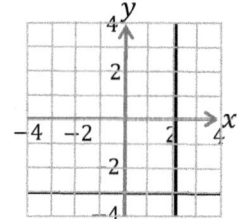

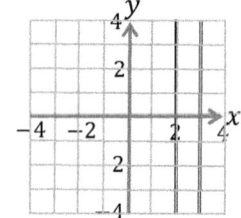

Module 3

Solid Figures

Family of Situations

At the end of module 3, the student is expected to be competent within the families of situations 'Usage of Technical Objects in everyday life'.

Categories of Action
The categories of action for module 3 include:
1. Recognition of objects,
2. Production of objects,
3. Determination of measures in the environment such as in school, at home, in the market place or on a journey.

Credit
The module is expected to be covered within 4 weeks teaching 4 hours per week (or within 15 to 16 hours).

Topic 14

Cubes and Cuboids

Objectives

At the end of this topic, the learner should be able to:

1. Identify, recognize and distinguish between cubes and cuboids.
2. Recognize the faces, vertices, edges of cubes and cuboids.
3. State the properties of cubes and cuboids.
4. Sketch cubes and cuboids.
5. Identify and recognize the faces and edges of cubes and cuboids.
6. Identify and recognize nets of cubes and cuboids.
7. Make models of cubes and cuboids from nets and use the various parts to establish the original figure.
8. Calculate the surface area and volume of cubes and cuboids.
9. Determine the capacity of a cubic container, a cuboid container, a hall, a bus, etc.
10. Establish the relationship between volume and capacity.

14.1 Vocabulary and Properties of Cubes and Cuboids

A cube and a cuboid are solid figures that look very alike. What two properties make them different?

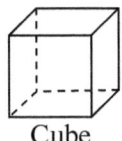
Cube

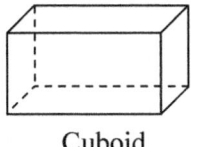
Cuboid

Faces, Edges and Vertices of Cubes and Cuboids

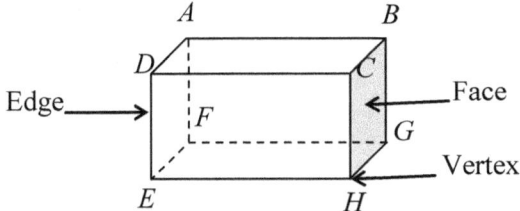

A **face** of a cube or cuboid is one of the flat surfaces.

An **edge** of a cube or cuboid is one of the lines on which any of the two faces meet.

A vertex is one of the corners at which any three faces or edges meet.

Notice how the vertices of a cube or cuboid are named using capital letters, the Edges are named with the two letters of the vertices that contain the edge and the faces are named with the four letters of the vertices that contain the faces.

A cube and a cuboid are both solid figures each having 6 (flat) plane faces, 8 vertices and 12 edges.

The faces of the above cuboid are *ABCD, EFGH, ADEF, BCHG, ABGF* and *CDEF*.

The edges of the above cuboid are *AB, BC, CD, AD, EF, FG, GH, HE, DE, AF, CH* and *BG*.

The vertices of the above cuboid are *A, B, C, D, E, F, G* and *H*.

The difference between a cuboid and cube is that each of the six faces of a cube is a square and all the 12 edges are equal. For a cuboid at least two opposite sides are rectangular.

From the above investigation, we can see that for a cube or cuboid;

Number of edges − Number of faces − Number of vertices = 2

14.2 Nets of Cubes and Cuboids

Suppose that someone wanted to make a cube out of a cardboard, he may sketch and cut out shapes similar to those in figure (a) and (b) below and fold along the dotted lines. He can then hold the meeting edges together using sticker tape.

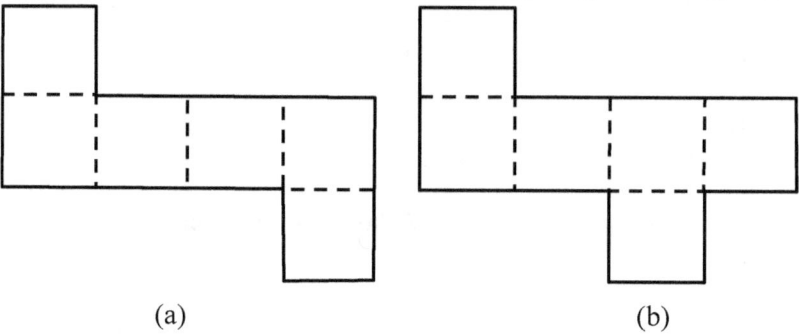

(a)   (b)

Exercise 14:1

1. How many faces has   (a) a cube?   (b) a cuboid?
2. How many edges has   (a) a cube?   (b) a cuboid?
3. How many vertices has (a) a cube?   (b) a cuboid?
4. Which of the following nets can you use to make a cuboid?

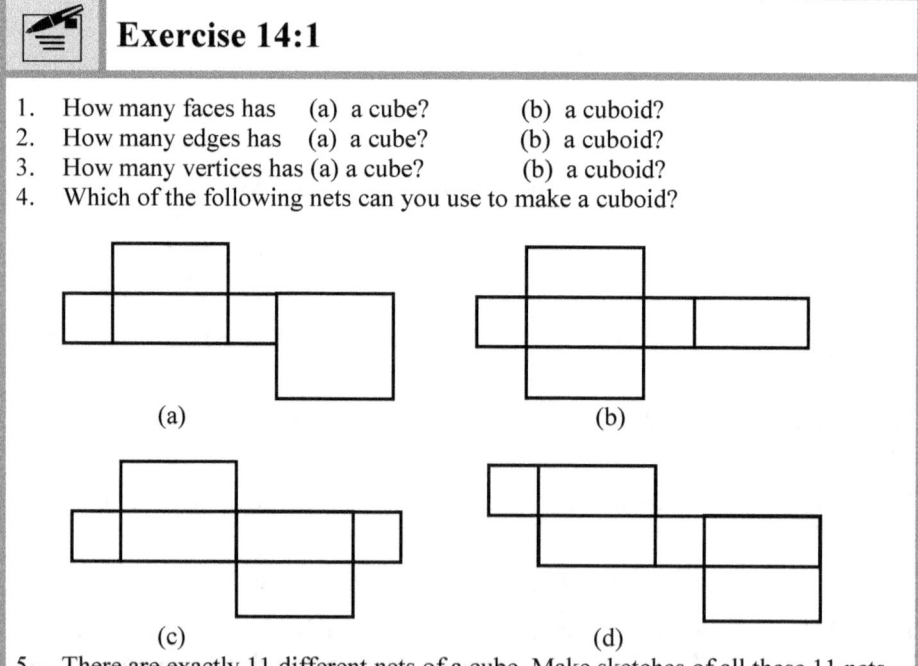

(a)   (b)

(c)   (d)

5. There are exactly 11 different nets of a cube. Make sketches of all these 11 nets.
6. Make five different nets of a cuboid and used each to make a cuboid.
7. List five containers or objects which are (a) cubes   (b) cuboids.

14.3 Surface Area of Cubes and Cuboids

 Integration Activity

The following figure is a net of a cuboid with length 8 cm, width 5cm and height 3 cm.

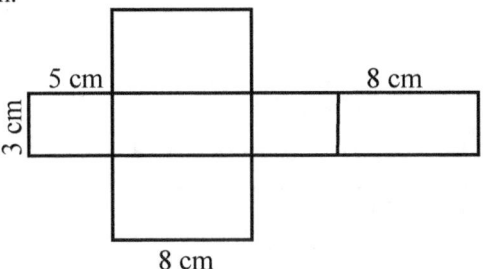

1. Describe the faces of this cuboid. Explain how you would use the area of the net to find the area of the cuboid.
2. Using the properties of a cube extend your ideas from (1) to find the area of a cube whose side is 9 cm.

From the net, it is clear that the cuboid is made of three pairs of opposite faces. Let the total surface area be TSA, then

$\therefore TSA$ = Sum of the areas of the three pair of opposite faces.

$\Rightarrow TSA = 2(3 \times 5) + 2(8 \times 5) + 2(3 \times 8) = 79 \text{ cm}^2$

A cube is a special cuboid with the length, width and height equal. Hence, it is even easier to calculate the total surface area by calculating the area of one face and multiplying by 6.

Hence if the length of a side is l units, then $TSA = 6(l \times l) = 6l^2$.

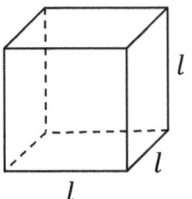

For a cube of side 9 cm, $TSA = 6l^2 = 6 \times 9^2 = 486 \text{ cm}^2$

Examples

1. The sides of a rectangular block are 5 cm, 11 cm and 40 cm. Calculate it surface area.

Solution

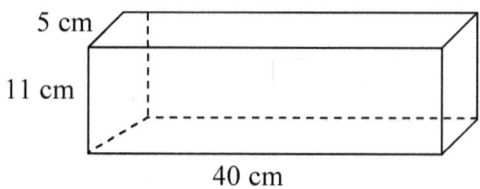

$$TSA = 2(5 \times 11 + 5 \times 40 + 11 \times 40) = 2(695) = 1390 \text{ cm}^2$$

2. A cube has side 9 cm. Find its total surface area.

Solution

$$TSA = 2(9 \times 9 + 9 \times 9 + 9 \times 9) = 2(243) = 486 \text{ cm}^2$$

14.4 Volume of Cubes and Cuboids

The volume of a solid is the number of cubic units that the solid contains. The basic unit of volume is the cubic centimeter (cm^3).

A cubic centimeter is defined as a cube whose side is 1 cm^3.

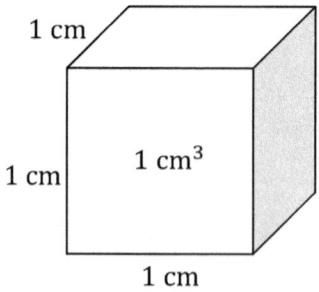

Investigative Activity

Requirements: Blocks such as magi cubes or match boxes.

1. Pack 3 layers of blocks so that each layer contains 4 rows of 6 blocks.
2. Count the number of blocks. How many blocks are there altogether?
3. Pack 2 similar layers on top of the first. Your blocks should lock something like in the following figure.

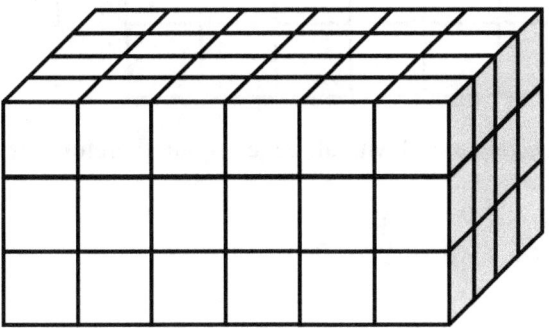

4. Remove the blocks one after the other and count them.
5. How many blocks are there altogether?
6. Now multiply 3 by 4 and by 6.
7. What conclusion do you draw?

Clearly, there are 3 layers and each layer contains 4 rows of 6 blocks (24 blocks). Therefore there are 72 blocks altogether.

Suppose that each small block has a volume of 1 cm^3, the total volume of the 72 blocks will be 72 cm^3.

Appreciate that if the blocks have side 1 cm^3 then by packing them as above, a cuboid, whose length, width and height are 6 cm, 4 cm and 3 cm respectively is obtained as in the figure below.

By multiplying the length, width and height the result is 72 cm^3. This suggest that

Volume of cuboid = length × width × height   or $V = lwh$

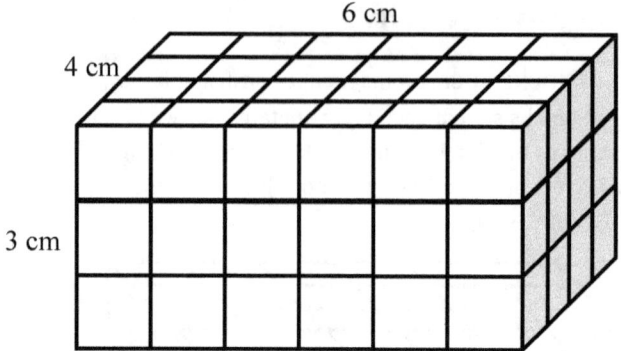

Since a cube is a special cuboid with all edges equal, therefore for a cube of side *l*,

Volume of cube $= l \times l \times l$ or $V = l^3$

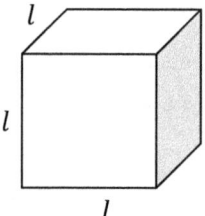

 Examples

1. The sides of a rectangular block are 5 cm, 11 cm and 40 cm. Calculate the volume of the block.

    Solution
    $V = lwh = 5 \times 11 \times 10 = 2100 \text{ cm}^3$

2. The length, width and height of a classroom are 7 m, 5 m and 4 m respectively. What is the volume of the classroom?

    Solution
    Volume of classroom $= lwh = (7)(5)(4) = 140 \text{ m}^2$

3. A cube has side 9 cm. Find its volume.

    Solution
    $V = a^3 = (9)^3 = 729 \text{ cm}^3$

Exercise 14:2

1. A box is 35 cm long, 20 cm wide and 12 cm high. Calculate the volume of the box.
2. What volume of sand will fill a room of length 15 m, breadth 12 m and height 8 m?
3. A box has a volume of 140,000 cm^3. If its breadth is 50 cm, and its length is 70 cm, find its height.
4. The volume of a rectangular tank is 520 m^3. Find the length of the tank given that the width and the height are 8 m and 5 m respectively.
5. Find the level that 270 m^3 of water will rise in a rectangular tank which is 6 m by 4.5 m.
6. Find the volume of a rectangular block 50 cm long and cross-sectional dimension of 3 cm by 20 cm.
7. Calculate the number of liters of water, which a rectangular tank with dimensions 8m by 6m by 3 m will hold.

14.5 Capacity

 Discussion Exercise

1. What do the following statements head in everyday life suggest about the meaning of the word capacity?
    (a) The hall is full to capacity.
    (b) The capacity of this car is four persons.
    (c) The capacity of this jar is 4 liters.
2. The capacity of a room or car or some container may vary. Why do you think this is so?
3. Volume sometimes influences capacity. Do you agree?

The **capacity** of a container is a measure of the space, which the container can accommodate. For cases (a) and (b), capacity means the maximum number of people, which the car or the hall can take. In this case, the unit of measurement is the number of seats, or the number of people. For case, (c) capacity means the amount of fluid (air or liquid), which a container can hold. Here the unit of measurement will be liters. Following this disparity, it is clear that if a room is 48 cubic meters by volume, air can occupy the whole space in the room. People on the other hand cannot occupy the whole volume of the room. The purpose for which people have to occupy a room is a factor, which will determine the capacity of the room. From the above discussion, it is clear that the volume of a container is always greater than or equal to its capacity. i. e. $V \geq C$.

The following are some units used to measure fluid capacity.

1 liter (l) = 1000 cm^3
1 cubic meter (m^3) = 1,000,000 cm^3 = 1000 liters
1 cubic centimeter (cm^3) = 1000 mm^3 = 0.001 liter

 Exercise 14:3

1. A container can take 3 liters of water. How much water can four such containers take? Give your answer in cm^3.
2. 3000 cm^3 of water is poured into a container, which already contained 7 liters of water. What is the total amount of water in the container? Give your answer in liters.
3. 9000 cm^3 of oil are removed from a tin, which originally contained 20 liters of oil. How much oil in cm^3 is left in the tin?
4. 24 women share a quantity of oil. If each has 3 liters what is the total quantity of oil in cm^3?
5. Convert to cm^3 (a) 7l (b) 40l (c) 30ml (d) 17.4 ml
6. Without measuring, pour into a bucket an amount of water which is about
   (a) 3 liters   (b) Half a liter.
7. How many liters of water will fill a rectangular tank of 500 cm by 100 cm by 200 cm?
8. Find the capacity in liters of a cuboid tank with dimensions 3 m by 4m by 5 m.
9. A cube measures 2 cm. Calculate
   (a) The volume of the cube   (b) The total surface area of the cube

 Multiple Choice Exercise 14

1. The shape of each side of a cuboid is;
   [A] A triangle   [B] A trapezium   [C] A circle   [D] A rectangle
2. A cuboid has ....vertices.
   [A] 4   [B] 6   [C] 8   [D] 12
3. A cuboid has ....faces.
   [A] 4   [B] 6   [C] 8   [D] 12
4. A cuboid has ....edges.
   [A] 4   [B] 6   [C] 8   [D] 12
5. The total surface area of a cube of edge 3 cm is:
   [A] 27 cm^2 [B] 27 cm^3 [C] 54 cm^2 [D] 36 cm^2
6. A rectangular tank, 2.25 m long and 1.6 m wide contains 2800 litres of water. Correct to the nearest cm, the depth of water in the tank is: (1000 cm^3 = 1 litre).
   [A] 778 cm   [B] 788 cm   [C] 770 cm   [D] 780 cm

Topic 15

Cylinders and Cones

Objectives

At the end of this topic, the learner should be able to:

1. Identify, recognize, describe and sketch cylinders and cones.
2. Distinguish between cylinders and cones.
3. Identify and recognize the base, lateral faces, lateral edge and altitude of a cylinder.
4. Identify and recognize the height, slant height, vertex, lateral surface and base of cones.
5. State the properties of cylinders and cones.
6. Identify and recognize nets of cylinders and cones.
7. Make models of cylinders and cones from nets and use the various parts to establish the original figure.
8. Calculate the surface area and volume of cylinders and cones.
9. Determine the capacity of a cylindrical container such as a drum or a conic container such a funnel etc.

CYLINDERS

15.1 Vocabulary and Properties of Cylinders

A right cylinder is a solid with two flat surfaces formed by two circles and a curved surface. The two circles are parallel and can be superimposed on each other.

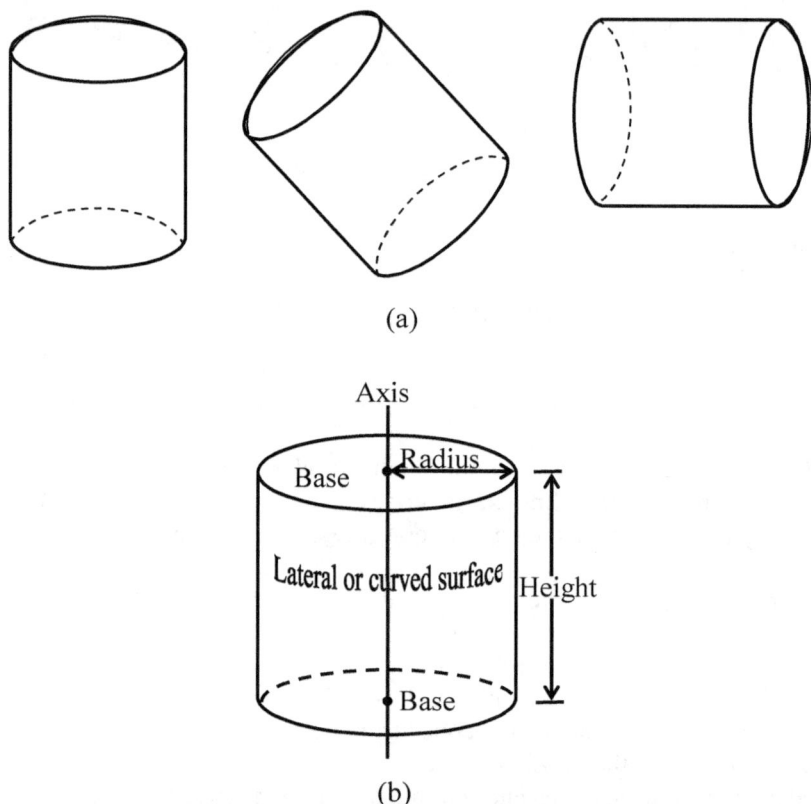

(a)

(b)

The two circles are the **bases** of the cylinder. The radius of the circles is the same as the **radius** of the cylinder. In a right cylinder, the distance between their centers is the **height** of the cylinder. The **lateral** or **curved surface** is shaped like a pipe with zero thickness. Figure (a) shows right circular cylinders drawn in perspective with dotted lines to show the hidden edges. Figure (b) is a right circular cylinder showing it various parts. The word **'right'** means that any plane drawn through the axis of the cylinder must be perpendicular to the base. Examples of right circular cylinders are a can, an unused piece of chalk, a milk tin, a cigarette etc.

Not all cylinders are right cylinders. The following figure shows an oblique cylinder. An oblique cylinder can be made by cutting a right cylinder diagonally on both sides.

15.2 Making a Circular Cylinder

To make a circular cylinder first trace it net on a cardboard and cut it out with a blade or scissors. Secondly, fold the net to obtain the circular cylinder.

 Integration Activity

Construct a cylinder with base radius 3.5 cm and height 10 cm.

Procedure

The two bases of the cylinder are circles, each with a radius 3.5 cm. The net for the curved surface is rectangular. Its dimensions are equal to the height of the cylinder (10 cm), and the circumference of the bases. Recall that the circumference C of a circle with radius r is given by $C = 2\pi r$. Thus,

$C = 2 \left(\frac{22}{7}\right) (3.5)$ cm $= 22$ cm

The following shows the net for the cylinder.

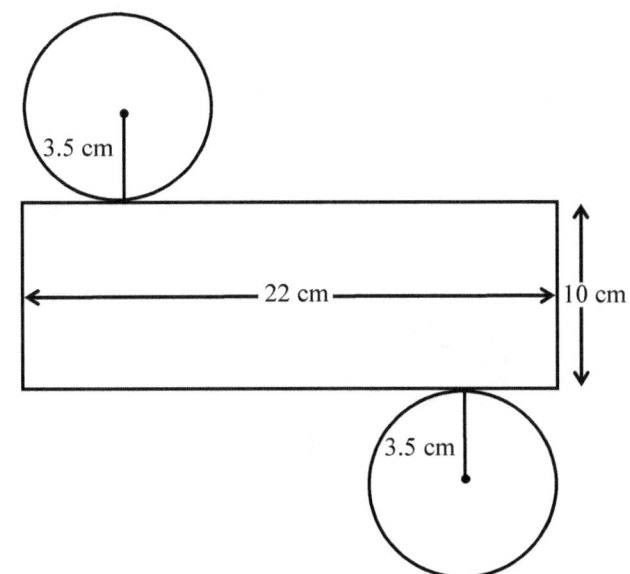

15.3 Surface Area of a Right Cylinder

The total surface area (TSA) of a cylinder is the same as the area of its net. A cylinder closed at both ends, is described as a **solid** cylinder. A cylinder open at both ends, is described as a **hollow** cylinder (or a pipe).

Therefore, there are three cases to consider when finding the surface area of a cylinder.

Surface area of hollow cylinder = Area of lateral surface

Surface area of cylinder open at one end = Area of lateral surface + Area of base

Surface area of solid cylinder = Area of lateral surface + 2× Area of base

 Example

1. Calculate the total surface of the right-circular solid cylinder that was constructed above.

   Solution
   Look again at the net of this cylinder. From this net,

   Total Surface Area = Area of circular bases + Area of lateral surface

   Area of circular base = $\pi r^2 = \frac{22}{7}(3.5)(3.5) = 38.5$ cm^2

   Area of both circular bases = 2×38.5 cm^2 = 77 cm^2

   Area of lateral surface = length × width = $22 \times 10 = 220$ cm^2

   Total Surface Area = 77 cm^2 + 220 cm^2 = 297 cm^2

2. An open drum has height of 30 cm and a radius of 7 cm. Calculate its surface area.

   Solution

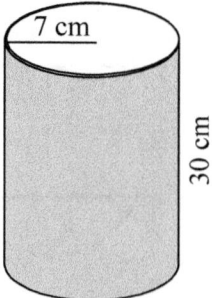

7 cm

30 cm

   Total Surface Area = Area of circular bases + Area of lateral surface

Area of circular base $= \pi r^2 = \frac{22}{7}(7)(7) = 154$ cm^2

Area of lateral surface $=$ Circumference $\times$ height

Circumference$= 2\pi r = 2\left(\frac{22}{7}\right)(7) = 44$

$\therefore$ Area of lateral surface $= 44 \times 30 = 1320$ cm^2

Total Surface Area $= 154$ cm$^2 + 1320$ cm$^2 = 1474$ cm^2

15.4 Volume of a Right Cylinder

It was earlier mentioned that a right circular cylinder is a prism because it has a uniform cross-section. Therefore, to find the volume of a cylinder multiply the base area by the height.

Volume of cylinder $=$ base area $\times$ height or $V = Ah$

Where A and h are the cross-sectional area and height of the prism respectively. Since the cross-section of a right circular cylinder is circular, for any right circular cylinder with radius r and height h its cross-sectional area A is given by $A = \pi r^2$. Hence the volume of the will be given by

$$V = \pi r^2 h$$

 Example

An open drum has height of 30 cm and a radius of 7 cm. Calculate its volume.

Solution

$$\text{Volume} = \text{base area} \times \text{ height}$$

$$\text{Base area} = \pi r^2 = \left(\frac{22}{7}\right)(7)^2 = 154 \text{ cm}^3$$

$$\text{Volume} = 154 \times 30 = 4620 \text{ cm}^3$$

243

 ## Exercise 15:1

In this exercise, take $\pi = \frac{22}{7}$ where necessary.

1. Water is full in a hollow cylinder of internal radius 10.5 cm and height 60 cm. A solid cone having the same height and radius is placed fully into the cylinder. Find the volume of water that will be left in the cylinder.

2. The following table shows the dimensions of cylinders. Complete the table

(a)	Base radius	5 cm	10 cm
(b)	Height	20 cm	30 cm
(c)	Surface area		
(d)	Volume		

3. A cylindrical tank 3.5 m in diameter contains water to a depth of 4 m. Find the total area of the wetted surface of the tank.

4. A Cuboid of sides' 2 cm by 4 cm by 11 cm is full of water. If this water is poured into a cylindrical jar of diameter 8 cm, find the depth of the water.

5. Calculate the volume of a cylindrical container with diameter 14 cm and height 7 cm.

6. The volume of a cylinder is 396 cm^3. Calculate the radius of the cylinder given that the height is 14 cm.

7. Find the height of a cylinder with radius 7 cm if its volume is 770 cm^3.

8. A drum is 20 cm in diameter and 70 cm tall. Calculate the capacity of the drum.

9. How many liters of liquid can a cylindrical can 14 cm in diameter and 20 cm high contain?

10. 90 liters of water is poured into a cylindrical bucket, which is 30 cm in diameter. Find the depth of water in the bucket.

11. The height of a cylinder with radius 35 cm is 21 cm. Find the curved surface area of the cylinder.

12. Find the area of the curved surface of a cylinder whose radius is 7 cm and whose height is 5 cm.

CONES

15.5 Vocabulary and Properties of Cones

A cone is a solid with a circular base that narrows to a point at the other end. Examples of conical shapes are ice cream, a funnel, a light beam from a flash light etc. The following are diagrams of right circular cones dawn in perspective.

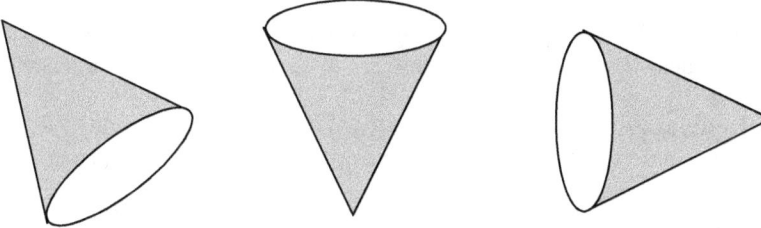

Below is a labeled diagram of the parts of a cone.

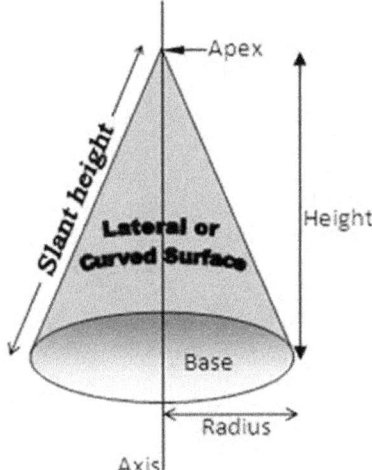

The circular end is called the **base**. The pointed (sharp) end is called the **apex**. The distance from the apex to the center of the base is called the **height**. The line of greatest slope from the apex to the circumference of the base is called the **slant height**. The **radius** is the distance from the center to the circumference of the base. The straight line that passes through the apex and the center of the base is called the axis.

15.6 Making a Cone

The net of a cone consist of a sector of a circle whose radius is equal to the slant height of the cone and a circular base whose circumference is equal to the length of the arc of the circle.

 Investigative Activity

Construct a cone whose lateral surface area is equal to $\frac{1}{3}$ the area of a circle of radius 5 cm and whose base radius is equal to 3 cm. What deductions do you make?

Requirements
Cardboard, scissors or blade, protractor, pair of compass, pencil cello tape.

Procedure
1. On a cardboard, use a pair of compass to draw a circle of radius 5 cm and cut it out with a scissors or blade.
2. With a pair of compass draw a circle of radius 3 cm and cut it out with a scissors or blade.
3. Fold the sector and join it with sticker tape along OA and OB.
4. Now close the open base with the disc you made in (2) and use sticker tape to keep them together.

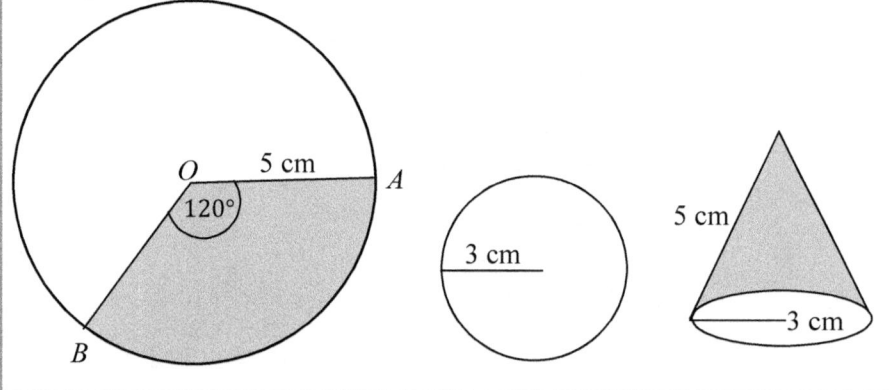

From above we see that:

(a) The slant height l of a cone is equal to the radius of the sector from which the cone is made.
(b) The lateral surface area of a cone is equal to the area of the sector from which the cone is made.
(c) The area of the base of a cone of radius r is equal to πr^2.

15.7 Surface Area of a Right Cone

Though the prove is beyond the scope of this book, we can show that the lateral surface area of a cone of radius r and slant height l is given by

$$S = \pi r l$$

The total surface area of the cone is given by

$S =$ Area of circular base $+$ lateral surface area

$$\Rightarrow S = \pi r^2 + \pi r l$$

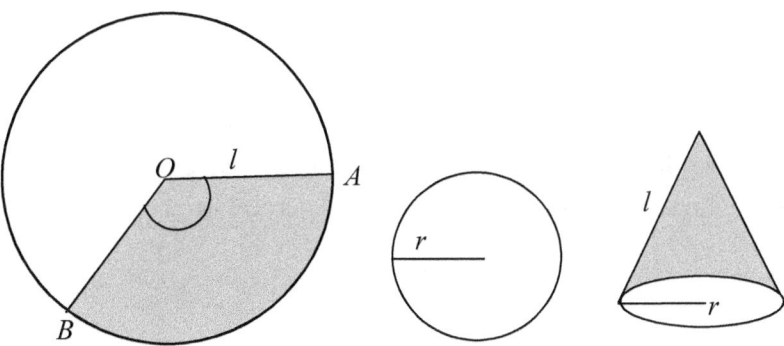

 Example

Calculate the total surface area of a right circular cone with base radius 3 cm and slant height 5 cm.

Solution

$S =$ Area of circular base $+$ lateral surface area

Area of circular base $= \pi r^2 = \frac{22}{7}(3^2) = \frac{198}{7}$ cm^2

Lateral surface area $= \pi r l = \frac{22}{7}(3)(5) = \frac{330}{7}$ cm^2

Total surface area of the cone , $S = \frac{198}{7} + \frac{330}{7} = 75.4$ cm^2

15.8 Volume of a Right Cone

 Investigative Activity

1. Construct an open base right circular cone with base radius 3 cm and height 4 cm using cardboard.
2. Construct a right circular cylinder open at one end with base radius 3 cm and height 4 cm using cardboard as shown below.
3. Fill sand in the cone right to the brim and pour it into the cylinder.
4. How many full cones fill the cylinder?

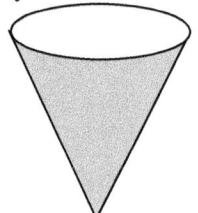 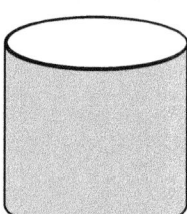

From the above investigation we see that the volume of any cone is equal to $\frac{1}{3}$ the volume of a cylinder with the same base area and vertical height. Hence

$$V = \frac{1}{3}Ah$$

Since base area $A = \pi r^2$, the volume of a right circular cone is given by

$$V = \frac{1}{3}\pi r^2 h$$

 Example

Calculate the volume of a cone whose height is 12 cm and whose base radius is 7 cm.

Solution

$V = \frac{1}{3}\pi r^2 h = \frac{1}{3}\left(\frac{22}{7}\right)(7)^2(12) = 616 \text{ cm}^3$

Exercise 15:2

1. Find the surface area of a cone whose base radius is 7 cm and whose slant height is 14 cm.
2. Find the surface area of a solid cone of radius 7 cm and slant height 9 cm.
3. The radius of a cone is 6 cm. Given that the height of the cone is 14 cm, calculate the volume of the cone.
4. Calculate the surface area of a cone with radius 14 cm and slant height 20 cm, when the base is; (a) open   (b) closed
5. Find to the nearest whole number the volume of a cone with radius 5 cm and height 12 cm
6. A cone has a diameter of 7 cm and a height of 12 cm. Calculate the volume of the cone.
7. Calculate the volume of a cone with base diameter 10 cm and height 12 cm.

 Multiple Choice Exercise 15

In the following exercises, where necessary, take $\pi = \frac{22}{7}$

1. A cylindrical container has a radius of 7 cm and a height 5 cm. The lateral surface area of the container is:
[A] 154 cm^2 [B] 220 cm^2 [C] 528 cm^2 [D] 770 cm^2
2. A cylindrical container has a radius of 7 cm and a height 5 cm. The volume of the container is:
[A] 154 cm^3 [B] 220 cm^3 [C] 528 cm^3 [D] 770 cm^3
3. The curved surface area of a cylindrical tin is 704 cm^2. when the radius is 8 cm the height is:
[A] 3.5 cm   [B] 7 cm   [C] 14 cm   [D] 28 cm
4. Correct to 1 decimal place the volume of a cylinder of height 8 cm and base radius 3 cm is: $\left(\text{Take } \pi = 3.142 \right)$
[A] 300.0 cm^3 [B] 250.0 cm^3 [C] 226.2 cm^3 [D] 150.9 cm^3
5. A solid cylinder of radius 7 cm is 10 cm long. Its total surface area is:
[A] 210 cm^2 [B] 594 cm^2 [C] 660 cm^2 [D] 748 cm^2
6. The internal and external radii of a cylindrical bronze pipe are 1.5 cm and 2 cm respectively. If the pipe is 10 cm long, the volume of the bronze used is
[A] $5\frac{1}{2}$ cm^3 [B] 55 cm^3 [C] $196\frac{2}{5}$ cm^3 [D] 550 cm^3
7. A cylinder 10 cm high is to be made with no overlap from the following a rectangular sheet of thin metal. The radius of this cylinder will be:

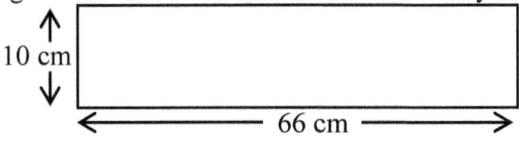

[A] 3.3 cm   [B] 6.6 cm   [C] 10.5 cm   [D] 21 cm

8. A cone of radius 3.5 cm is made from a sector of a circle of radius 14 cm. The area of the curved surface of the cone is:

    [A] 22 cm^2 [B] 88 cm^2 [C] 77 cm^2 [D] 154 cm^2

9. The volume of a cone of radius 3.5 cm and vertical height 12 cm is:

    [A] 15.5 cm^3 [B] 21.0 cm^3 [C] 42.0 cm^3 [D] 154.0 cm^3

10. A sector is cut off from a circle of radius 8.2 cm to form a cone. If the radius of the resulting cone is 3.5 cm, then, the curved surface area of the cone is:

    [A] 12.83 cm^2 [B] 22.0 cm^2 [C] 67.2 cm^2 [D] 90.2 cm^2

11. A cone is 14 cm deep and the base radius is $4\frac{1}{2}$ cm. The volume of water that is exactly half the volume of the cone is

    [A] 49.5 cm^3 [B] 99 cm^3 [C] 148.5 cm^3 [D] 297 cm^3

12. The total surface area of a solid circular cone with base radius 3 cm and slant height 4 cm is:

    [A] 66 cm^2 [B] $75\frac{3}{7}$ cm^2 [C] $78\frac{2}{7}$ cm^2 [D] 88 cm^2

Module 4

Elementary Statistics

Family of Situations

At the end of module 4, the student is expected to be competent within the families of situations 'Organization of information and estimation of quantities in the consumption of goods and services'.

Categories of Action

The categories of action for module 4 include:
1. The collection, organization and exploitation of information
2. The interpretation of results.

Credit

The module is expected to be covered within 3 weeks teaching 4 hours per week (or within 10 to 12 hours).

Topic 16

Simple Data, Collection and Representation

Objectives

At the end of this topic, the learner should be able to:

1. Explain the meaning of statistics and state the importance of statistics in real life and for studies.
2. Collect raw data within the environment using a variety of different methods.
3. Choose the best method for the collection of a given data.
4. Distinguish between discrete and continuous data.
5. Tally and tabulate the data collected.
6. Represent data on frequency tables and statistical graphs such as pictograph, bar charts and pie charts
7. Interpret statistical graphs such as pictograph, bar charts and pie charts.
8. Determine the mode, median and mean of a small range of data considering, odd and even number of scores for median.

16.1 Notion of Statistics

Information such as shoe sizes, ages of people, number of students in each class of a school and so on, is known as **data**. The branch of mathematics, which deals with the collection, processing and analyzing of data, is called **statistics**, though the word statistics is sometimes used as a synonym for data.

16.2 Importance of Statistics

Statistics enables people to make predictions and take proper decisions. For instance in order to make sure that the General Certificate of Education questions are distributed to examination centers in such a way that excesses are minimized and shortages are avoided, the number of candidates who will go in for a particular subject in each center, must be known. This is the main reason why during registration every student indicates the subjects he will write.

16.3 Data Collection

There are many methods of collecting data and some methods are more suitable in some situations than others. The methods of collecting data include:

Census Method

In this method, we count the population and other related items directly.

Questionnaire Method

A questionnaire is a set of well-structured questions used to gather information from a sample of the population.

Survey Method

This method requires the use of tools or experiments to observe and record the required data.

Interviews Method

This oral method involves asking questions to each individual of the sample and recording their responses. The method is often used as a last tool in selecting students into a school or workers for a job.

Research or Archive Method

Prerecorded information can be obtained from books, newspapers, information storage centers or the internet.

The method used in each case depends on the nature and purpose of the data, the tools available, the skill of the data collector and many other factors.

 Group Activity

In groups of eight to ten students carry on the following activity.
(1) As a group, collect the following information concerning members of your class.
   (a) Division of origin (b) sex (c) age   (d) height   (e) shoe size
   (f) Favourite fruit of student   (g) witches and wizards in the class
(2) The teacher solves 10 simple problems on familiar topics on the board. Each group of students' record the number of problems solved in intervals of 1 minute until the problems are finished.
(3) Write a report concerning your data collection exercise in (1) above. In your report include the difficulties you encountered.

 Discussion Exercise

1. Explain the difficulties you encountered in the course of collecting the data in the group activity above.
2. Justify the authenticity of your results in each case.

16.4 Qualitative and Quantitative Data

There are two types of data-qualitative and quantitative data. Data that can be expressed numerically is called quantitative data. Examples of such data are; the number of students of age 13 years, the marks obtained by students in a test etc. Data that cannot be expressed numerically is called qualitative data. Examples of such data are the names of students in a class, the division of origin of each of the students in form one etc.

 Exercise 16:1

1. During an interview of which you are a participant, you are given the following form to fill. You may not cancel any thing on this form.
   (a) Fill the form carefully and classify the data as qualitative or quantitative.

Data	Individual information	Type of data
Name of student		
Class of student		
Name of guardian		
Village of origin		
Center Number		
Candidate name		
Candidate number		
Sex of candidate		
Age of candidate		
Religion of candidate		
Fee paid by candidate		

    (b) What method is being used here to collect the information?

2. What method would you use to collect the following data about each of your classmates? In some cases there may be more than one method.
    (a) Name of student (b) Age of student (c) Weight of student
    (d) height of student (e) Sex of student (f) Fee paid by student
    (g) Favourite colour of student   (h) Shoe size of student.
    (i) Previous school attended by student (j) Blood group of student.
    (k) Number of hours which the student spent per day reading.
    (l) Official language spoken by student.

3. Copy the following table. Collect the data asked for concerning 5 of your classmates and record on the table.

Name of classmate	Age	Height	Shoe size	Previous school

4. Classify the following as qualitative or quantitative data, giving reasons for your answer.
    (i) Temperatures over a given period (ii) Rainfall in Bamenda in 2011
    (iii) Names of your family members (iv) Ages of your classmates

5. The following table shows the enrolment of a certain school for six years from 2005 to 2011.

Year	2006	2007	2008	2009	2010	2011
Enrolment	401	403	444	485	527	586

Use the data to answer the following questions.
    (a) State the difference in the population between 2006 and 2011.
    (b) Is the school enrolment increasing?
    (c) Given that, there are about 40 students per class. How many classes are there in the school?
    (d) In which year is the school likely to have started a new form one stream?
    (e) In which year will that form one stream be in form five?
    (f) Estimate the enrolment of the school in 2012

16.5 Discrete and Continuous Data

Discrete data is data which consists of exact and separate distinct values. Data such as the number of objects is discrete because it can take only the values 0,1,2,3 .... Data such as the suit of playing cards is discrete because it can take only the values clubs, diamonds, hearts and spades. Other examples of discrete data include center number, sex, country of origin, year of birth, class of student etc.

Continuous data is data which does not consist of exact and separate or distinct values but consist of a finite or infinite interval. Data such as the heights or weight of objects is continuous because between any two values of such data we can insert other values. For instance in the interval [1, 2], we can still insert values such as 0.1, 0.2, 0.3 etc. In the interval [0.2, 0.3], we can still insert values such as 0.21, 0.22, 0.23, 0.29, etc.

16.6 Raw Data and Data Processing

Whichever method is used to collect data, unprocessed and assorted data called **raw data** is usually obtained. To properly appreciate data it is necessary to analyze or process the data.

An example of raw data is the data which you collected during the previous group work.

Data procession requires two main steps and methods.

(i) Tallying

In tallying strokes are made against each statistic in such a way that each stroke represents one element of a specific class. The fifth stroke is made horizontal across the first four. For instance, if a statistic occurs seven times, the tallying will be ⊪ II. Each stroke is called a **tally** and this method of processing raw data is known as the **tally method**.

(ii) Frequency-Distribution Tables

The number of times an item, x occurs is called **frequency**, usually denoted by f. A **frequency-distribution table** shows the frequency f of each item x, and the process of drawing the table is known as **tabulation**.

 Exercise 16:2

Tally the data for shoe sizes of children of your class and draw a frequency-distribution table of the data.

Your answer in the above exercise may look something like this.

Tally for shoe sizes

Shoe size	Tally
37	II
38	II
39	HHH I
40	HHH I
41	IIII
42	HHH II
43	HHH I
44	IIII
45	I

Frequency-distribution table

Shoe size, x	Frequency, f
37	2
38	2
39	6
40	6
41	4
42	7
43	6
44	4
45	1

To crosscheck the information, add the frequency (number of shoes). The sum should correspond to the total (number of students in your class). If not rectify it, by repeating the tallying or the survey.

Exercise 16:3

1. A boy measured to the nearest meter, how far he could through a tennis ball and obtained the following results.

    66   69   70   68   71   68   69   70   67   68
    67   68   67   66   69   68   69   70   68   67

    Tally the raw data and hence draw a frequency-distribution table of the data.

2. The following are the marks gained by 30 students in an examination.

57	60	58	62	56	59
62	60	62	63	64	56
60	62	63	58	59	63
66	66	56	58	51	58
53	57	59	57	53	54
49	54	60	64	60	64

    Tally the marks and use it to draw a frequency-distribution table of the data.

3. In a certain restaurant in Bamenda town, rice(r), beans (b), plantains (p) and yams(y) are sold. The following shows the food ordered by the customers of the restaurant on a certain day.

p	y	b	r	y	p	r	y	b	b	r	y	p	p	y
b	r	y	p	p	y	p	y	b	r	y	p	r	y	b
p	y	p	y	y	b	y	b	r	y	p	b	r	y	p

Tally the data and draw a frequency-distribution table which would be used to determine the number of customers that ordered each of the food.

4. The weights of 60 students were measured and recorded as follows.

61	57	62	60	55	66	63	64	52	58
64	59	64	57	62	61	61	55	62	60
64	61	64	57	62	59	64	69	62	64
63	64	66	56	62	66	58	63	58	58
60	60	64	55	60	63	58	58	66	58
60	63	61	59	66	63	63	61	58	64

Tally the weights and draw a frequency- distribution table of the data.

16.7 Representation of Data-Statistical Graphs

The various ways of representing data include pictograms, bar charts, pie charts, histograms etc. The names 'chart', 'graph', or 'diagram' are synonymous and shall be used in this topic interchangeably.

(a) Pictograms or Ideographs

Pictures used to represent data in such a way that one can see at a glance the relative frequency (or the frequency of one data item with respect to another), of any of the item. Pictograms are used for comparative data such as population of countries.

 Exercise 16:4

Using a pictogram. Represent the data you collected concerning the favourite fruit of students in your class in the previous group activity.

Your answer should look something like the following which represents20 pineapples, 40 Bananas, 50 watermelons and 90 Oranges.

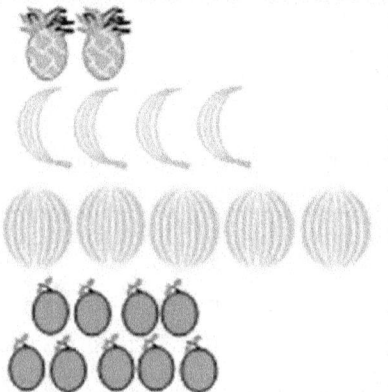

1 Fruit represents10 fruits

(b) Pie Charts or Circular Diagrams

In this type of graphs, a circle, divided into sectors is used to represent the frequencies. Each sector is proportional to the statistic representing it. It is therefore necessary that pie charts should always be drawn with a pair of compasses and the angles measured with a protractor. To draw a pie chart the angles must first be calculated.

 Exercise 16:5

Using a pie chart represent the data you collected concerning the favourite fruit of students in your class in the previous group activity.

Your answer should look something like the following which represents 20 pineapples, 40 Bananas, 50 watermelons and 90 Oranges.

Type of fruit	Number of fruits	Calculation	Angle representing fruit
Pineapples(P)	200	$\frac{200}{2000} \times 360$	36°
Bananas(B)	400	$\frac{400}{2000} \times 360$	72°
Watermelons(W)	500	$\frac{500}{2000} \times 360$	90°
Oranges(O)	900	$\frac{900}{2000} \times 360$	162°
TOTAL	2000		360°

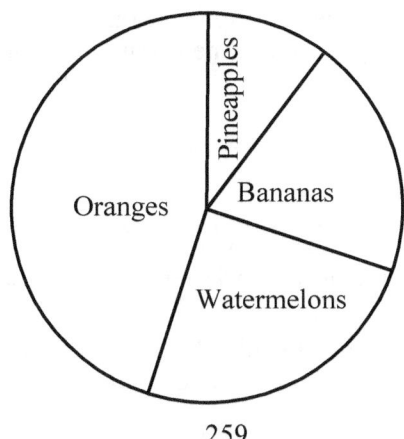

(c) Bar Charts

In this case, bars of equal width represent information. Only the length or height of the bars has any significance. Therefore, the width of the bars can be any convenient size. The bars may be together or separated but must not overlap. Bar charts are of many types and are usually though they could be horizontal.

 Exercise 16:6

Using a bar chart represent the data you collected concerning the favourite fruit of students in your class in the previous group activity.

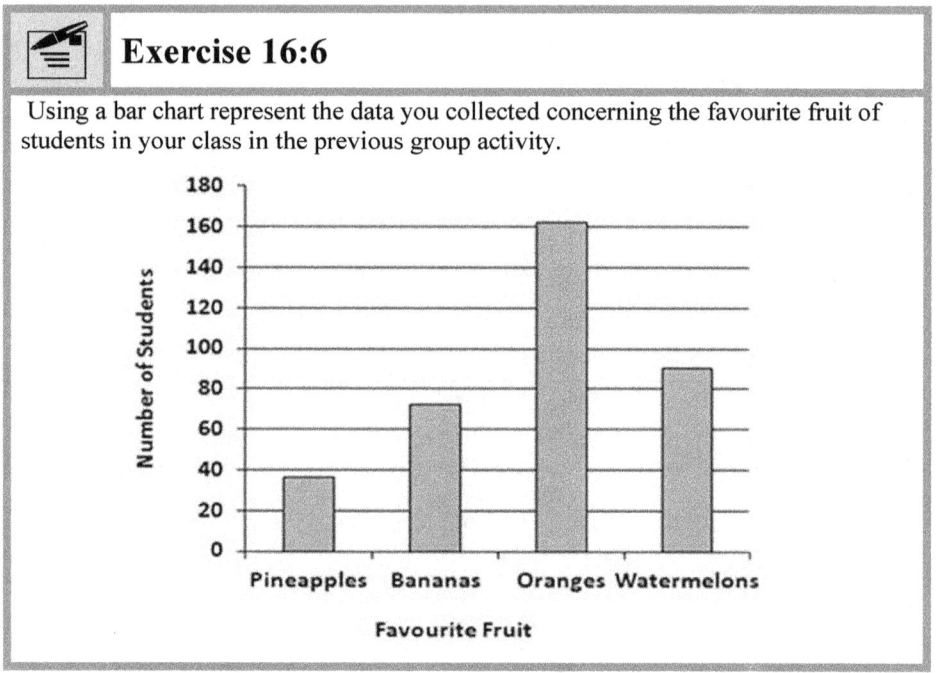

(d) *The proportionate bar*

In the proportionate bar chart, the bars are of equal widths and placed one upon another. When the bars are placed one upon the other the height of the bars is proportional to the frequency otherwise the width of the bars is proportional to the frequency.

 Exercise 16:7

Using a proportionate bar chart, represent the data you collected concerning the favourite fruit of students in your class in the previous group activity.

Your answer should look something like the following which represents20 pineapples, 40 Bananas, 50 watermelons and 90 Oranges.

| Watermelons, 25% |
| Oranges, 45% |
| Bananas, 20% |
| Pineapples, 10% |

(e) Chronological Bar Chart

This type of bar chart usually represents the variation of some statistic over a period.

 Exercise 16:8

Using a Chronological bar chart, represent the number of problems your teacher solved over the period during the group activity above.

Your answer should look something like the following chart which is constructed for the following table.

Number of problems	1	2	3	4	5
Time in minutes	1	3	5	8	10

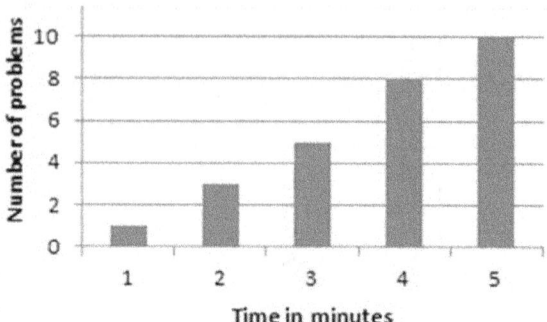

 Exercise 16:9

1. The pictogram shows the work force of the Fruit industry, the Cattle industry and the Electricity Company in Cameroon in a certain year.
   (a) Which of the pictures represents the fruit industry?
   (b) Which of the institutions has the greatest number of workers?
   (c) Which of the institutions has the least number of workers?
   (d) Does the fact that there are seven fruits mean that there are only seven workers in the fruit industry?

   (i)

   (ii)

   (iii)

2. During a Youth Week, the form three students of a certain school participated in team sports as shown in the following table.

Team sport	Number of students
Handball	45
Basketball	60
Football	75

Represent this information on a pie chart and state the angle for basketball.

3. The pie chart above (not drawn to scale) shows how a student spent his pocket money amounting to 27,000 FCFA. (The angles are in degrees.) Given that he spent twice as much on books as he did on taxi, calculate
   (a) How much he spent on books (b) How much he spent on others.

4.

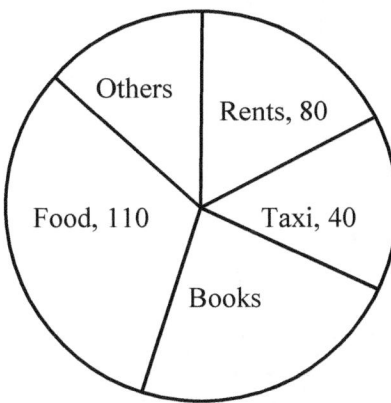

5. The table below is a survey carried out on a group of students to find out what they ate for launch on a certain day. Draw a histogram to display this data.

Achu	15
Rice	9
Garri	4
Bread	2

6. The pie chart below shows the number of votes for candidates A, B and C in an election. Calculate the percentage of the votes to the nearest whole number in favour of candidate B.

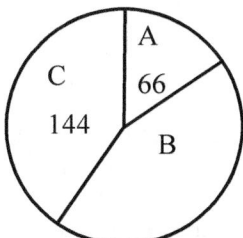

7. The livestock of a certain farm consist of 28 cows, 300 sheep, 74 pigs, 306 poultry, 9 dogs and 3 cats. If this information is recorded on a pie chart, calculate

the angle in degrees, at the center of the sector representing the cows.

8. Five boys A, B, C, D and E are of heights 160, 144, 120, 96 and 80 centimeters respectively. Represent this information on a bar chart.

9. The figure below shows a pie chart indicating the favourite colours of a group of 108 girls.
   (a) Find the angle of the sector for girls who like yellow.
   (b) Find the number of girls who like green.

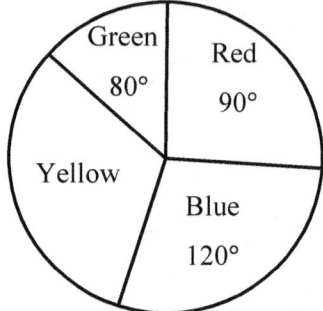

10. The following table was used to draw a pie chart. Find the values of x and y.

Item type	A	B	C	D
Frequency	48	88	y	32
Sector angle (°)	72	x	108	48

✍ Multiple Choice Exercise 16

1. Information collected in form of figures is called;
   [A] Census [B] Data [C] Population [D] Sample

2. What number does the following tally marks represent?
   ЦЦ ЦЦ ЦЦ ЦЦ ЦЦ III
   [A] 18 [B] 23 [C] 28 [D] 33

3. Which of the following is the correct tally representation of 17 students in a class?
   [A] ЦЦ ЦЦ ЦЦ III [B] ЦЦ ЦЦ IIIIIII [C] IIIII IIIII IIIII III [D] ЦЦ ЦЦ ЦЦ II

4. When recording data, the tally marks ЦЦ ЦЦ ЦЦ III, will be recorded as;
   [A] 13 [B] 15 [C] 18 [D] 20

5. In a school examination, 480 candidates scored Grade 4 out of 720 candidates. What would be the angle at the center of a pie chart, showing all the grades for grade 4?
   [A] 270° [B] 240° [C] 210° [D] 180°

6. A pie chart is drawn to represent the following percentages: 20%, 50%, 25% and

5%. The angle which represents 5% is

[A] 5°   [B] 18°   [C] 25°   [D] 126°

7. The data below shows the frequency distribution of marks scored by a group of students in a class test. The number of students who took the test is

[A] 14   [B] 15   [C] 18   [D] 20

Marks	2	3	4	5	6
Frequency	2	4	5	3	1

8. The table below shows the scores of 15 students in a physics test. The number of students who scored at least 5, is

[A] 6   [B] 8   [C] 9   [D] 7

Marks	1	2	3	4	5	6	7	8	9	10
No. of students	1	3	2	0	1	6	1	0	1	0

9. The distribution by province of 840 students in the faculty of science of the University of Buea in a certain session is as follows:

Adamawa Province	45
North West Province	410
Littoral Province	105
West Province	126
South West Province	154

In a pie chart drawn to represent this distribution, the angle subtended by West Province is:

[A] 42°   [B] 45°   [C] 48°   [D] 54°

10. The following pie chart represents the fruits on display in a grocery shop. If there are 60 oranges in display, the number of apples are

[A] 40   [B] 80   [C] 90   [D] 120

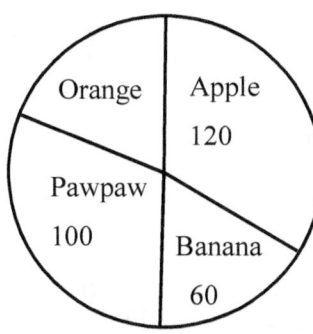

Answers to Structural Exercises

Exercise 1:1

1. Groups of ten.

2.(a) 9 ΛΛΛ||||
ΛΛΛΛ|||||

(b) 𐄂𐄂99 ΛΛΛ||||
𐄂Λ99ΛΛΛ||||

(c) 𐄂𐄂𐄂99ΛΛ|
𐄂𐄂𐄂9ΛΛ

(d) 99Λ|
99Λ

(e) 𐄂99ΛΛ|||
𐄂99ΛΛΛ||

(f) 𐄂99 Λ||
9ΛΛ||

(g) Λ||||
||||

(h) 𐄂9999ΛΛΛΛ||||
9999ΛΛΛΛ||||

(i) 𐄂𐄂99ΛΛΛΛ|||
𐄂𐄂99ΛΛΛ|||||

(j) 9ΛΛΛΛ|||
9ΛΛΛΛ|||

3. (a) 42   (b) 3207   (c) 2142   (d) 2428   (e) 1161

   (f) 1334   (g) 734   (h) 6742   (i) 5523   (j) 16545

4. (a) 𐄂9999ΛΛ|
999 ΛΛΛ||

(b) 𐄂9ΛΛΛΛ||
𐄂9Λ ΛΛΛ|||

5. 𐄂𐄂𐄂9999ΛΛΛ|||||
𐄂𐄂𐄂9999ΛΛΛΛ|||

Exercise 1:2

1. (a) XXIX   (b) XLVIII   (c) MMMMDCCCLXXIV
   (d) MMMCMXCIII   (e) MCCCXXXVIII   (f) MCMXC
   (g) MMM   (h) MMCDXXXII   (i) DXLIX
   (j) MCDXXVIII   (k) MMMMCDLXXXVIII (l) CCLXXXVI

2. (a) One hundred and fifty four
   (b) Six hundred and nine
   (c) One thousand one hundred and fourteen
   (d) Seven hundred and sixty one
   (e) Five hundred and eighty nine
   (f) One thousand nine hundred ninety six
   (g) Fifty nine

266

(h) One thousand four hundred.
(i) Three thousand four hundred and forty.
(j) Eight hundred and one
(k) Two thousand four hundred and fifty
(l) Two thousand four hundred.
3. I, II, III, IV, V, VI, VII, VIII, IX, X, XI, XII, XIII, XIV, XV, XVI, XVII, XVIII, XIX, XX, XXI, XXII, XXIII, XXIV, XXV, XXVI, XXVII, XXVIII, XXIX, XXX, XXXI, XXXII, XXXIII, XXXIV, XXXV, XXXVI, XXXVII, XXXVIII, XXXIX, XL, XLI, XLII, XLIII, XLIV, XLV, XLVI, XLVII, XLVIII, XLIX, L
4. (a) DXXII   (b) MMCCXXIII   (c) MMMMDVII
    (d) MCCCXXXIV (e) XLIV   (f) CCCXXV

5. (a) ₹999ΛΛΛΙΙΙ / 9999ΛΛΙΙΙΙΙ   (b) ΛΛΛΛΙΙΙΙ / ΛΛΛΛΙΙΙ   (c) ΛΛΙΙΙ / ΛΛΙΙ

    (d) 999ΛΛΛΙΙΙ / 999ΛΛΛΙΙ   (e) ₹9ΛΛΛΙ / ₹9ΛΛΛΛ   (f) ₹9999ΛΛΛΛΛΙΙΙ / 99999ΛΛΛΛΙΙΙΙΙ

Exercise 1:3

1. (a) 30 thousand   (b) 3 thousand   (c) 3 hundred thousand

2. 68 stands for six tens, eight units while 86 stands for eight tens, six units.

3. (a) Five thousand five hundred and seventy eight

   (b) Fifty thousand four hundred and forty eight

   (c) Eight hundred and ninety three thousand two hundred and sixty one
   (d) seventeen thousand two hundred and four.
4. (a) 538001   (b) 17004   (c) 9909   (d) 232
   (e) 111101   (f) 8080   (g) 10010

Skill Building Exercise 2:1

1. (a) $30676 < 60793 < 75077 < 98706$

   (b) $98706 > 75077 > 60793 > 30676$

2. (a) $3 < 7$ (b) $10 > 8$ (c) $14 > 13$ (d) $4 < 6$

   (e) $5 - 1 < 9$ (f) $26 - 3 > 5$ (g) $6 - 3 > 2$ (h) $4 + 2 < 8$

   (i) $2 < 7 - 4$ (j) $11 < 5 + 8$ (k) $3 \times 4 > 2 + 7$ (l) $18 \div 6 < 7$

   (m) $458 < 764$ (n) $963 > 687$ (o) $8463 > 8379$ (p) $6536 < 8543$

3. $56 < 68 < 74 < 121.$ 4. $901 > 847 > 585 > 497$

Skill Building Exercise 2:2

1. (a) 90350   (b) 111052   (c) 553752   2. 278.

Skill Building Exercise 2:3

(1) 110   (2) 30   (3) 97   (4) 131   (5) 28   (6) 76

(7) 40   (8) 103   (9) 124   (10) 49   (11) 83   (12) 71

Skill Building Exercise 2:4

1. (a) 2^7 (b) 7^5 (c) 3^9 (d) 6^5 (e) 8^8

2. (a) 9×9 (b) $5 \times 5 \times 5$ (c) $7 \times 7 \times 7 \times 7 \times 7 \times 7$

   (d) $2 \times 2 \times 2 \times 2 \times 2 \times 2 \times 2 \times 2$ (e) $3 \times 3 \times 3 \times 3$

   (f) 10×10 (g) $10 \times 10 \times 10 \times 10 \times 10$ (h) $10 \times 10 \times 10 \times 10$

3. (a) 81   (b) 125   (c) 117649   (d) 256   (e) 81

   (f) 100   (g) 100000   (h) 10000

4. (a) 10^5 (b) 10^7 (c) 10^8

Exercise 2:1

1. (a) Yes   (b) No   (c) Yes   (d) No   (e) Yes   (f) Yes
   (g) Yes   (h) No$_t$ (i) Yes   (j) Yes   (k) No   (l) Yes

Skill Building Exercise 2:5

1. (a) $8 \times 10 + 7$ (b) $1 \times 10^2 + 2 \times 10 + 4$

   (c) $3 \times 10^3 + 2 \times 10^2 + 4$ (d) $6 \times 10^3 + 7$

   (e) $9 \times 10^4 + 6 \times 10^3 + 8 \times 10^2$ (f) $7 \times 10^6 + 5 \times 10^4 + 3 \times 10^2$

2. (a) 46   (b) 752   (c) 8307   (d) 3005   (e) 16300   (f) 4060200

Skill Building Exercise 2:6

1. (i) $3 \times 5^3 + 2 \times 5^2 + 1 \times 5^1 + 4$

   (ii) $1 \times 2^6 + 1 \times 2^4 + 1 \times 2^3 + 1 \times 2^2 + 1$
   (iii) $5 \times 7^4 + 2 \times 7^3 + 3 \times 7^2 + 1$ (iv) $4 \times 6^4 + 3 \times 6^3 + 2 \times 6 + 1$
   (v) $5 \times 8^4 + 3 \times 8 + 2$ (vi) $2 \times 7^4 + 4 \times 7^3 + 2$
   (vii) $2 \times 3^4 + 1 \times 3^2 + 2 \times 3 + 1$ (viii) $3 \times 8^4 + 4 \times 8^3 + 2 \times 8 + 1$
   (ix) $7 \times 9^3 + 4 \times 9^2 + 8 \times 9$ (x) $1 \times 2^3 + 1 \times 2^2 + 1$
   (xi) $4 \times 9^3 + 8 \times 9^2 + 3 \times 9 + 6$ (xii) $2 \times 4^3 + 3 \times 4^2 + 1$
   (xiii) 4×6^2 (xiv) $7 \times 8^3 + 1 \times 8^2 + 4$
   (xv) $1 \times 4^4 + 2 \times 4^2 + 2 \times 4 + 1$ (xvi) $2 \times 3^3 + 1 \times 3^2$

2. (a) 32_{four} (b) 571_{eight} (c) 214_{five}

(d) 6003_{seven} (e) 43100_{six} (f) 3060800_{nine}

Skill Building Exercise 2:7
(1) 47   (2) 59   (3) 167   (4) 961
(5) 412   (6) 93   (7) 1703   (8) 11635
(9) 27   (10) 2590   (11) 127   (12) 147
(13) 5041   (14) 1399   (15) 1846   (16) 353029

Skill Building Exercise 2:8
(1) 1100010101_{two} (2) 11443_{eight} (3) 1244_{six} (4) 233301_{four}
(5) 1211222_{three} (6) 656_{nine} (7) 11101_{two} (8) 102523_{six}
(9) 1430_{eight} (10) 10663_{seven}

Skill Building Exercise 2:9
1. 65_{seven} 2. 33_{eight} 3. 333020_{five} 4. 2131_{four} 5. 21112201_{three}
6. 145_{nine} 7. 5252_{seven} 8. 4051_{six} 9. 1123111_{four} 10. 122_{seven}
11. 662_{eight} 12. 501_{six} 13. 1201323_{four} 14. 113_{nine} 15. 12000021_{three}

Skill Building Exercise 2:10
1. (a) 100010_{two} (b) 111001_{two} (c) 101110_{two}
   (d) 111111_{two} (e) 1001000_{two}
2. (a) 53   (b) 27   (c) 109   (d) 85   (e) 29
3. (a) 1000_{two} (b) 1000001_{two} (c) 101100_{two} (d) 101001_{two}
   (e) 11111_{two} (f) 11101_{two} (g) 11110_{two} (h) 100100110_{two}
   (i) 10011100_{two} (j) 10_{two} (k) 10_{two} (l) 100_{two}

Skill Building Exercises 3:1
1. (a) negative 7   (b) positive 13   (c) positive 10   (d) negative 31
   (e) positive 3 plus positive 7   (f) negative 5 minus negative 2
   (g) negative 10 plus negative 2   (h) positive 8 minus positive 10
   (i) positive 14 + negative 1)   (j) negative 9 minus positive 6
2. (a) 30 francs gained   (b) 50 francs lost
   (c) 10 francs lost   (d) 20 francs gained
3. (a) 12 points down (b) 6 points up (c) 7 points down (d) 9 points up
4. (a) 6 days ago (b) 2 days ahead (c) 4 days ago (d) 3 days to come.
5. (a) 144 m below sea level   (b) 27 m above sea level
   (c) 35 m above sea level   (d) 17 m below sea level
6. (a) 10 francs spent (b) 35 francs spent (c) 20 francs saved (d) 200 francs saved

7. (a) 0   (b) +7   (c) −9   (d) +8
8. (a) <   (b) <   (c) >   (d) <   (e) >   (d) =
9. −11, −14, −9, −3, +2, +3, +5, +13   10.   −5, −4, −2, +5, +7, +9

11.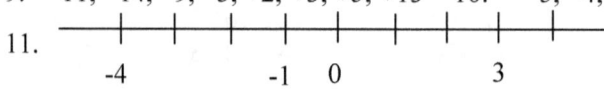

12. (a)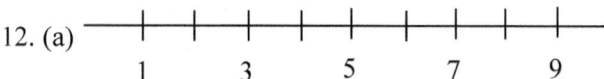

(b)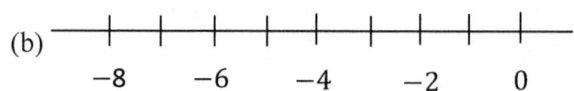

13. (a) has borrowed 17,000, is owing 17,000
    (b) has saved 13, 000, has accredit of 13,000
14. (a) 20   (b) 13   (c) 58   (d) 356
15. (a) <   (b) <   (c) >   (d) >
16. (a) true   (b) false   (c) true   (d) true

Skill Building Exercises 3:2

1. 13   **2.** −14   **3.** −20   **4.** −10   **5.** 28   **6.** 0   **7.** 20

8. −6   **9.** 25   **10.** 14   **11.** 19   **12.** −85   **13.** −3   **14.** 1

Skill Building Exercises 3:3

1. (a) − 25   (b) +53   (c) − 42   (d) + 74

2. (a) 27   (b) −20   (c) 10   (d) 20   (e) −12   (f) −16   (g) −10   (h) 4
   (i) 24   (j) −4   (k) 5   (l) 1   (m) 25   (n) 45

Skill Building Exercises 3:4

1. (a) − 6   (b) −12   (c) 15   (d) −30 (e) 16 (f) −64

2. (a) 15 (b) −6   (c) −6 (d) −40 (e) −8   (f) 40 (g) 24 (h) −72

Skill Building Exercises 3:5

1. −47   2. −11   3. 6   4. −75   5. 3   6. 67

Exercise 4:1
1.   (a)   9 Hours (b) 540 minutes (c)   32400 s
2. 144 days 3. 730 seconds 4. 65 minutes   5.   6.50 a.m.
6. (a) 28 hours (b) 1680 minutes (c) 100800 seconds

7. 443 days   8. 10080 minutes

Exercise 4:2

1. (a) 5:25 p.m. (b) 11:36 p.m.   (c) 2:20 p.m. (d) 11:10 a.m.

2. (a) 20:34 (b) 5:56   (c) 12:45 (d) 00:45

3. (a) 9 hours 30 minutes (b)14 hours 30 minutes

    (c) 19 hours 2 minutes (d) 10 hours 42 minutes

4. (a) ante meridiem (between midnight and midday)

    (b) post meridiem (between midday and midnight)

5.   3 a.m.   **6.** 19 hours 30 minutes   **7.** (a) 12:46 p.m. (b) Wednesday

Exercise 4:3

1. (a) $140°F$ (b) $95°F$ (c) $68°F$ (d) $104°F$ (e) $167°F$

2. (a) 50 °C   (b) 34 °C   (c) 17 °C   (d) 93 °C   (e) 45 °C

3. (a) hot (b) warm (c) cold (d) warm (e) cold (f) cold

    (g) hot (h) warm (i) cold (j) warm (k) cold (l) cold

Exercise 5:1

1. (i) $T = \{0, 1, 3, 6, 10, 15, 21, 28, 36, 45, 55\}$

    $S = \{0, 1, 4, 9, 16, 25, 36, 49\}$

    $R = \{0, 1, 2, 4, 6, 8, 9, 10, 12, 14, 15, 16, 18, 20, 21, 24, 26, 27, 28, 30, 32, 33, 34, 35,$
    $36, 38, 39, 40, 42, 44, 45, 46, 48, 49, 50, 51, 52, 54, 55, 56, 57, 58, 60\}$

(ii) (a) $\{0, 1, 4\}$ (b) $T = \{0, 1, 6\}$ (c) $\{0, 1\}$

          T_4 T_5

3.(a)   1 + 3 + 5 + 7 + 9 = 25

  (b)   0 + 1 + 2 + 3 + 4 + 5 = 10

  (c)   15 − 5 = 10

271

Exercise 5:2

1. (i) (a) $S = \{0, 4, 16, 36, 64, ...\}$, (b) $R = \{0, 2, 4, 6, 8, 10\}$

   (c) $T = \{0, 6, 10, 28, 36\}$

2. (a)
$$
\begin{array}{rcccc}
1 & - & 0 & = & 1 \\
4 & - & 1 & = & 3 \\
9 & - & 4 & = & 5 \\
16 & - & 9 & = & 7 \\
25 & - & 16 & = & 9 \\
36 & - & 25 & = & 11 \\
\end{array}
$$
   (b) Odd numbers

c. (i)
$$
\begin{array}{rcccc}
1 & + & 0 & = & 1 \\
3 & + & 1 & = & 4 \\
6 & + & 3 & = & 9 \\
10 & + & 6 & = & 16 \\
15 & + & 10 & = & 25 \\
\end{array}
$$
   (ii)
$$
\begin{array}{rcccc}
1 & - & 0 & = & 1 \\
3 & - & 1 & = & 2 \\
6 & - & 3 & = & 3 \\
10 & - & 6 & = & 4 \\
15 & - & 10 & = & 5 \\
\end{array}
$$

   (d) Square numbers   Natural numbers

Exercise 5:3

1. (a) $\{1,2,3,4,6,8,12,24\}$
   (b) $\{1,2,3,4,5,6,10,12,15,20,30\}$
   (c) $\{1,2,3,4,5,6,8,10,12,15,20,24,30,40,60,120\}$
   (d) $\{1,2,3,4,6,8,9,12,18,24,36,72\}$
   (e) $\{1,3,5,7,15,21,35,105\}$
   (f) $\{1,3,5,15,25,75\}$
2. (a) $\{2,4,6,8,10\}$ (b) $\{4,8,12,16,20\}$ (c) $\{5,10,15,20,25\}$
   (d) $\{7,14,21,28,35\}$ (e) $\{8,16,24,32,40\}$ (f) $\{12,24,36,48,60\}$
3. $\{1, 2, 3, 4, 6,8\}$ 4. $\{18, 24, 30,36\}$

Exercise 5:4

1. $\{5,29,47\}$.   2. $\{49,35,24\}$.

3. 15, $\{2,3,5,7,11,13,17,19,23,29,31,37,41,43,47\}$

4. $\{1,2,3,4,6,8,9,12,18,24,36,72\}$. 9 composite factors

Skill Building Exercises 5:1

1. (a) 2^5 (b) 3^4 (c) $2^2 \times 3 \times 5$ (d) $2^3 \times 3^2$ (e) $3^2 \times 5$

   (f) $3^2 \times 7$ (g) 3×17 (h) $2^4 \times 3$ (i) 3^5

2. (a) $2 \times 3^2 \times 5 \times 7$ (b) $2^3 \times 5^2 \times 11$ (c) $2 \times 7 \times 11 \times 13$

   (d) $2^6 \times 3^3$ (e) $2^5 \times 3 \times 5 \times 11$

Skill Building Exercises 5:2

(a) HCF = 6, LCM = 36   (b) HCF = 2, LCM = 144

(c) HCF = 24, LCM = 288   (d) HCF = 3, LCM = 168

(e) HCF = 4, LCM = 24   (f) HCF = 2, LCM = 24

(g) HCF = 6, LCM = 72   (h) HCF = 3, LCM = 105

(i) HCF = 5, LCM = 75   (j) HCF = 72, LCM = 4320

2. HCF = 3, LCM = 210

Skill Building Exercises 5:3

	(a)		(b)		(c)		(d)	
(1)	(a)	169	(b)	961	(c)	289	(d)	400
(2)	(a)	324	(b)	784	(c)	361	(d)	1024
(3)	(a)	28	(b)	17	(c)	80	(d)	68
(4)	(a)	25	(b)	23	(c)	32	(d)	53
(5)	(a)	64	(b)	343	(c)	2197	(d)	8000
(6)	(a)	125	(b)	1331	(c)	512	(d)	1000
(7)	6							

Skill Building Exercises 5:4

1. (a) 85564, 21342, 4378, 23490, 6936. The last digit is even.
   (b) 64665, 21342, 97965, 23490, 6936. The sum of digits is divisible by 3.
   (c) 85564, 6936. The number formed by the last two digits is divisible by 4.
   (d) 64665, 97965, 76445, 23490. The last digit is 5 or 0.
   (e) 23490, 6936. The number is even and sum of digits is divisible by 3.
   (f) 3689, 97965. The difference between twice the last digit and the rest of the number is 0 or is divisible by 7.
   (g) 6936. The number formed by last three digits is divisible by 8.
   (h) 64665, 97965, 23490. The sum of the digits is divisible by 9.
   (f) 23490. The last digit is 0.
   (j) 4378. The difference between the sum of the odd digits and the sum of the even digits is 0.
   (k) 6936. The sum of digits is divisible by 3 and number formed by last two digits is divisible by 4.

2.

Number	Divisible by										
	2	3	4	5	6	8	9	10	25	50	100
12644750	x			x				x	x	x	
74319275				x					x		
1861425		x		x			x		x		
6671456300	x		x	x				x	x	x	x
925675435				x							

Exercise 6:1

1. (a) 2   (b)   5   (c)   7   2. (a) 3   (b)   8   (c) 4
3 (a) four-thirds   (b) one and five-eighths   (c) seven tenth
   (d) two-thirds   (e) nine-fifth   (f) 5 and two-thirds
4. (a) $\frac{2}{3}$ (b) $3\frac{4}{5}$ (c) $\frac{16}{3}$ (d) $\frac{5}{9}$ (e) $\frac{4}{3}$ (f) $8\frac{1}{10}$
5. (a) improper   (b) mixed number   (c) proper
   (d) proper   (e) improper   (f) mixed number

Skill Building Exercise 6:2

1. (a) $\frac{60}{100}$ (b) $\frac{36}{24}$ (c) $\frac{25}{40}$ (d) $\frac{24}{36}$ (e) $\frac{28}{12}$ (f) $\frac{60}{135}$

2. (a) $\frac{11}{40}$ (b) $\frac{3}{2}$ (c) $\frac{4}{5}$ (d) $\frac{4}{3}$ (e) $\frac{4}{3}$ (f) $\frac{20}{27}$

Skill Building Exercise 6:2

1. (a) $\frac{8}{3}$ (b) $\frac{16}{5}$ (c) $\frac{7}{4}$ (d) $\frac{16}{5}$ (e) $\frac{13}{5}$

2. (a) $3\frac{2}{3}$ (b) $7\frac{1}{2}$ (c) $2\frac{1}{4}$ (d) $1\frac{4}{7}$ (e) $4\frac{1}{3}$

Skill Building Exercise 6:3

1. (a) $\frac{7}{9} > \frac{6}{9}$ (b) $\frac{2}{3} < \frac{4}{5}$ (c) $\frac{3}{6} \equiv \frac{2}{4}$ (d) $\frac{3}{4} < \frac{7}{9}$

2. $\frac{1}{4}, \frac{1}{3}, \frac{2}{3}, \frac{1}{2}, \frac{2}{3}, \frac{3}{4}, \frac{6}{7}, \frac{7}{8}, \frac{8}{9}$

3. (a) $\frac{2}{6}, \frac{3}{6}, \frac{4}{6}, \frac{5}{6}, \frac{7}{6}$ (b) $\frac{2}{8}, \frac{4}{8}, \frac{6}{8}, \frac{7}{8}$ (c) $\frac{9}{11}, \frac{9}{8}, \frac{9}{7}, \frac{9}{5}, \frac{9}{2}$ (d) $\frac{10}{5}, \frac{12}{4}, \frac{11}{3}, \frac{16}{4}$

274

4. (a) $\frac{17}{9} > 1\frac{6}{9}$ (b) $\frac{32}{30} > \frac{4}{5}$ (c) $\frac{33}{66} \equiv \frac{22}{44}$ (d) $1\frac{3}{4} < 1\frac{7}{9}$

5. $5\frac{1}{3}, 4\frac{2}{5}, \frac{13}{3}, 3\frac{1}{2}, \frac{20}{6}, 3\frac{1}{4}, \frac{16}{5}, \frac{18}{6}, \frac{17}{8}, 1\frac{2}{3}$

Skill Building Exercise 6:4

1. (a) $\frac{6}{7}$ (b) $\frac{10}{11}$ (c) $\frac{3}{2}$ (d) $\frac{5}{9}$ (e) $\frac{2}{5}$

2. (a) $\frac{2}{9}$ (b) $\frac{2}{7}$ (c) $\frac{1}{4}$ (d) $\frac{5}{13}$ (e) $\frac{2}{5}$

3. (a) $1\frac{5}{12}$ (b) $1\frac{19}{63}$ (c) $1\frac{9}{40}$ (d) $\frac{29}{30}$ (e) $\frac{7}{10}$

4. (a) $\frac{4}{9}$ (b) $\frac{20}{63}$ (c) $\frac{37}{88}$ (d) $\frac{24}{91}$ (e) $\frac{7}{20}$

5. (a) $1\frac{49}{60}$ (b) $1\frac{61}{63}$ (c) $1\frac{39}{40}$ (d) $1\frac{9}{30}$ (e) $1\frac{1}{5}$

Skill Building Exercise 6:5

1. (a) $15\frac{4}{5}$ (b) $10\frac{5}{7}$ (c) $11\frac{39}{40}$ (d) $12\frac{3}{10}$ (e) $8\frac{1}{5}$

2. (a) $7\frac{3}{5}$ (b) 1 (c) $4\frac{29}{40}$ (d) $\frac{11}{30}$ (e) $3\frac{1}{5}$

Real life Exercise page 93: $15\frac{27}{40}$

Skill Building Exercise 6:6

1. 40 2. 135 3. $\frac{15}{32}$ 4. $\frac{3}{8}$ 5. 6 6. 3 7. 6

8. 10 9. $\frac{6}{11}$ 10. $\frac{2}{15}$ 11. $\frac{7}{12}$ 12. $\frac{3}{4}$ 13. $\frac{1}{30}$ 14. $\frac{1}{20}$ 15. $\frac{3}{10}$ 16. 10 17. 14 18. 78 19. 30 20. $3\frac{1}{8}$ 21. $5\frac{5}{6}$ 22. $12\frac{2}{15}$ 23. 56 24. 56 25. $21\frac{2}{3}$ 26. $7\frac{3}{20}$

Skill Building Exercise 6:7

1. (a) $\frac{3}{7}$ (b) $-\frac{9}{2}$ (c) $\frac{4}{3}$ (d) $\frac{8}{5}$ (e) $-\frac{11}{6}$ 2. (a) 21 (b) $\frac{3}{20}$

3. (a) $\frac{1}{6}$ (b) $\frac{6}{7}$ (c) 1 (d) $\frac{5}{3}$ or $1\frac{2}{3}$ 4. (a) $\frac{7}{9}$ (b) $\frac{3}{4}$ (c) $\frac{3}{2}$

5. (a) $\frac{1}{9}$ (b) 36

Integration Activity Page 97: (i) $8\frac{1}{8}$ kg (ii) $14\frac{2}{13}$ kg

Skill Building Exercise 6:8

(a) $\frac{4}{5}$ (b) $\frac{7}{18}$ (c) $1\frac{13}{15}$ (d) $1\frac{5}{16}$

Real life Exercise page 100:   (a) $\frac{1}{14}$ (b) $\frac{3}{35}$

Integration Activity Page 101

1. (a) 120 (b) (ii) $\frac{7}{10}$ (iii) $\frac{1}{25}$ (c) $\frac{3}{7}$ **2.** The cable is short by $\frac{1}{6}$ m.   3. $207\frac{7}{8}$ m 4. Yes because its area is 68.90625 m² which is slightly less.

Skill Building Exercise 7:1

1. (a) 0.675 (b) 5.86   (c) 0.389896   (d) 0.76587   (e) 0.4535

2. (a) seven millionth   (b) twenty thousandth

   (c) two hundredth   (d) three thousandth

Skill Building Exercise 7:2

1. (a) 0.75 (b) 0.5 (c) 0.4 (d) 0.625 (e) 1.5 (f) 1.25 (g) 2.25 (h) 5.48
2. (a) $\frac{4}{5}$ (b) $\frac{13}{20}$ (c) $1\frac{23}{100}$ (d) $3\frac{3}{4}$ (e) $7\frac{1}{2}$ (f) $2\frac{8}{25}$

Skill Building Exercise 7:3

(1) 144.3 (2) 2.973 (3) 24.21 (4) 14.35   (5) 37.4958   (6) 176.7522 (7) 16.86 (8) 150.79 (9) 6.52 (10) 3.8585 (11) 6.1225 (12) 15.4

Skill Building Exercise 7:4

1. 1.35   2. 70.5042   3. 24.65991   4. 199.4905

5. 274.428   6. 0.197508   7. 0.2795   8. 230.7102

9. 33.07392   10. 1.20995   11. 0.048   12. 0.036

13. 6300   14. 934   15. 2750   16. 65

Skill Building Exercise 7:5

(1) 0.716   (2) 0.019   (3) 1.3   (4) 2.337   (5) 0.502   (6) 0.16   (7) 13   (8) 11.28 (9) 390   (10) 0.1036 (11) 1.52 (12) 0.2395 (13) 3.4   (14) 0.35

Skill Building Exercise 7:6

1. (a), (c), (d), (f), (h), (j), (l)

2. (a) 0.25 (b) 0.77$\overline{7}$ (c) 0.375 (d) 1.5 (e) 0.2727 (f) 0.44

 (g) 0.166 (h) 0.325 (i) 0.$\overline{571428}$ (j) 0.15 (k) 4.16$\overline{6}$ (l) 0.7

Skill Building Exercise 7:7

Express the following numbers in standard form.

(a) 5×10^3 (b) 4.8×10^2 (c) 1.02×10^4 (d) 7×10^5

(e) 3.2×10^{-3} (f) 7.3×10^{-5} (g) 9.25×10^{-1} (h) 1×10^{-3}

(i) 5.6×10^{-1} (j) 3×10^{-5} (k) 1.96×10^{-3} (l) 3.4×10^{-10}

Exercise 8:1

1. (a) $\dfrac{1}{4}$ (b) $\dfrac{18}{25}$ (c) $\dfrac{41}{50}$ (d) $\dfrac{19}{20}$ (e) $\dfrac{27}{200}$ (f) $\dfrac{1}{2000}$

 (g) $\dfrac{29}{400}$ (h) $\dfrac{139}{400}$ (i) 4 (j) $2\dfrac{1}{2}$ (k) 6 (l) $7\dfrac{1}{2}$

2. (a) 75% (b) 80% (c) 300% (d) 34% (e) 45% (f) 52%

Exercise 8:2

1. (a) 40% (b) 75% (c) 250% (d) 4035%

2. (a) 0.2 (b) 0.35 (c) 1.15 (d) 2.5

Exercise 8:3

(1) (a) 15 (b) 5527.5 (c) 5.6 (d) 81000

(2) 18 (3) 16% (4) 2.5% (5) 8% (6) $33\dfrac{1}{3}$%

Exercise 8:4

1. (a) 3:5 (b) 2:3 2.(a) 3:4 (b) 3:7 (c) 4:7

3. (a) 2:3 (b) 1:2 (c) 2:3 4. 800 Frs., 1200 Frs.

5. Bih =20 kg, Mankaa = 50 kg 6. 10,000 FCFA, 8000 FCFA, 6000 FCFA.

7. Ndi = 12, Shey = 18, Nfor = 24.

8. (a) 3375000 CFA (b) 1125000 CFA, 1500000 CFA

Exercise 9:1

1. (a) a ruler, a textbook, the walls of your classroom, a star.
 (b) an unused piece of chalk, the moon, the sun, a wire, a wheel.

2. (a) The length of a line *AB*.
  (b) A ray beginning from *B* and passing through *A*
  (c) A ray beginning from *A* and passing through *B*.
  (d) A line passing through the points *A* and *B*.
  (e) A line segment with end points *A* and *B*.
  (f) A line segment with end points *A* and *B*.
3. (a) 4 unit   (b) 12 units   (c) 8 units   (d) 16 units (e) 12 units
4. (a) [*AB*); A ray from *A*, passing through *B*.
  (b) [*PQ*]; A line segment with end points *P* and *Q*.
  (c) (*LM*); A line passing through the points *L* and *M*.
  (d) (*XY*]; A ray from *Y*, passing through *X*.
5. (a) 5   (b) 8 (c) 6 (d) 8 (e) 10 (f) 6 (g) 3 (h) 12

Exercise 9:2

1. 10; *ABC, DEFGH, IJK, ADI, CHK, AEJ, BEI, BGK, CGJ*.
2. (a) *CDE*   (b) answers vary.
  (c) [*DE*), [*DC*), [*DA*), [*DB*), (d) [*CD*], [*DE*], [*CE*], [*DA*], [*DB*]

Exercise 9:3

1. (a) 18 mm   (b) 44 mm   (c) 26 mm
3. (a) metres   (b) metres   (c) kilometres
  (d) centimeters   (e) centimetres or decimeters.

Exercise 10:1

1. (a) 30   (b) 60°   (c) 90°   (d) 120°   (e) 150°   (f) 180°
  (g) 210°   (h) 240°   (i) 270°   (j) 300°   (k) 330°   (l) 0° or 360°
2. (a) $\frac{1}{2}$ (b) $\frac{1}{4}$ (c) $\frac{1}{6}$ (d) $\frac{1}{12}$ (e) $\frac{1}{8}$ (f) $\frac{3}{8}$ (g) $\frac{5}{6}$ (h) $\frac{2}{3}$
3. (a) $\frac{1}{2}$ (b) $\frac{1}{3}$ (c) $\frac{1}{6}$ (d) $\frac{1}{4}$ 4. (a) $\frac{1}{8}$ (b) $\frac{1}{18}$ (c) $\frac{1}{10}$ (d) $\frac{1}{60}$
5. (a) 180°   (b) 90°   (c) 60°   (d) 165°
6. (a) 360°   (b) 420°   (c) 300°   (d) 180°
7. (a) 0°   (b) 90°   (c) 180°   (d) 120°
8. (a) acute   (b) reflex (c) obtuse (d) acute   (e) straight (f) reflex
9. Acute: b, e, i   Obtuse: a, f, h, j, l   Straight: d
  Reflex: c, k,   Revolution: g

Exercise 10:2

1. (a) $p + q = 180°$ (b) $x = y$
    (c) $x + y + w + z = 360°$ (d) $f = 45°$
2. (a) $x = 40°$ (b) $p = 75°, q = 105°$ (c) $y = 110°, z = 70°$
    (d) $m = 89°$ (e) $v = 37°$ (f) $n = 82°$ (g) $w = 92°$
    (h) $s = 135°, t = 142°$ (i) $u = 60°, v = 80°$ (j) $k = 124°$
3. x is greater than 270° but less than 360°

4. (i) $a = 100°$, $b = 80°$ (ii) $x = 60°$ **5.** $a + b + c = 180°$

Exercise 10:3
(a) $45°$ (b) $69°$ (c) $96°$ (d) $221°$

Exercise 11:1
1. (a) trapezium   (b) rectangle   (c) parallelogram   (d) rhombus   (e) square

2.

	Property	Square	Rectangle	Rhombus	Parallelogram	Kite	Trapezium
a	All the sides are equal	Y	N	Y	N	N	N
b	All the diagonals are equal	Y	Y	N	N	N	N
c	Diagonals bisect each other	Y	Y	Y	Y	Y	N
d	Diagonals are perpendicular	Y	N	Y	N	Y	N
e	Diagonals bisect opposite angles	Y	Y	Y	N	Y	N
f	Adjacent sides are equal	Y	N	Y	N	Y	N
g	Opposite sides are equal and parallel	Y	Y	Y	Y	N	N
h	Only two sides are parallel	N	N	N	N	N	Y
i	Adjacent sides are perpendicular	Y	Y	N	N	N	N

3. $x = 110°$, $y = 110°$

Exercise 11:2

1. (a) 16 m^2 (b) 10 m^2 (c) 18 m^2 2. 84.81 km^2 3. 17.3 km^2

Exercise 11:3

1. (i) (a) 1750 m^2 (b) 170 m (ii) (a) 5400 m^2 (b) 300 m
  (iii) (a) 512 m^2 (b) 96 m (iv) (a) 3000 m^2 (b) 70 m
2. 4   3. 26.3   4. $45\,000 \text{ m}^2$ 5. 5   6. 15 m, 54 m   7. 11 cm   8. 26 m
9. (i) (a) 2601 cm^2 (b) 204 cm (ii) (a) 144 cm^2 (b) 48 cm
  (iii) (a) 225 cm^2 (b) 60 cm (iv) (a) 900 cm^2 (b) 120 cm
10. 56 cm   11. (a) 289 cm^2 (b) 68 cm

Exercise 11:4

1. 40 cm^2 2. 60 cm^2 3. (a) 96 cm^2 (b) 10 cm   4. (a) 48 cm   (b) 864 cm^2
5. 21 cm   6. 6 cm   7. (a) 10 cm   (b) 42 cm^2 8. (a) $\sqrt{41}$ cm   (b) 38 cm^2 9.
126 cm^2 10. 46 cm^2, $4\sqrt{2}$ cm   11. 7 cm.   12. 12 m^2

Exercise 11:5

1. (a) scalene triangle   (b) isosceles triangle   (c) equilateral triangle
   (d) equilateral triangle (e) scalene triangle   (f) isosceles triangle

2. (a) right-angled triangle (b) obtuse angle triangle (c) acute angle triangle

3. (a) isosceles triangle (b) right-angled triangle   (c) acute angle triangle

5. (a) false   (b) false   (c) true   (d) true   (e) true   (f) true

6. (a) isosceles triangle   (b) scalene triangle   (c) scalene triangle
   (d) equilateral triangle

7. (a) obtuse angle triangle (b) right-angled triangle (c) acute angle triangle

Exercise 11:6

(a) $x = 70°$ (b) $x = 72°, y = 54°$ (c) $y = 100°$ (d) $y = 50°, x = 70°$

Exercise 11:7

1. (a) 30 cm^2 2. 6 cm 3. cm^2 4. (a) 6 cm (b) 4 cm (c) 12 cm^2

Exercise 11:10

1. (i) (a) chord   (b) radius   (c) diameter   (d) minor arc
       (e) secant   (f) tangent   (g) major arc
   (ii) (a) semi-circle   (b) minor sector   (c) minor segment

2. (a) Minor arc   (b) radius   (c) isosceles
   (d) chord   (e) Minor sector

3. (a) Minor segment (b) Minor arc (c) radius
   (d) diameter   (e) Chord   (f) Minor sector

Exercise 11:10

1. 38.5 cm^2 2. 154 cm^2 3. 22 cm 4. 88 cm

5. 7 cm 6. 1886.5 cm^2 7. 132 m 8. 115.5 cm^2

Exercise 12:1

1.

Name of figure	Diagram	No. of lines of symmetry
(a) An isosceles trapezium		1
(b) A kite		1
(c) An isosceles triangle		1
(d) A rhombus		2
(e) A rectangle		2
(f) An equilateral triangle		3
(g) A parallelogram		0
(h) A square		4
(i) A circle		Infinite diametrical lines of symmetry

2. A, B, C, D, E, K, M, T, U, V, W, Y   3. H, I, O, X

4. F, G, J, L, N, P, Q, R, S, Z,

5.

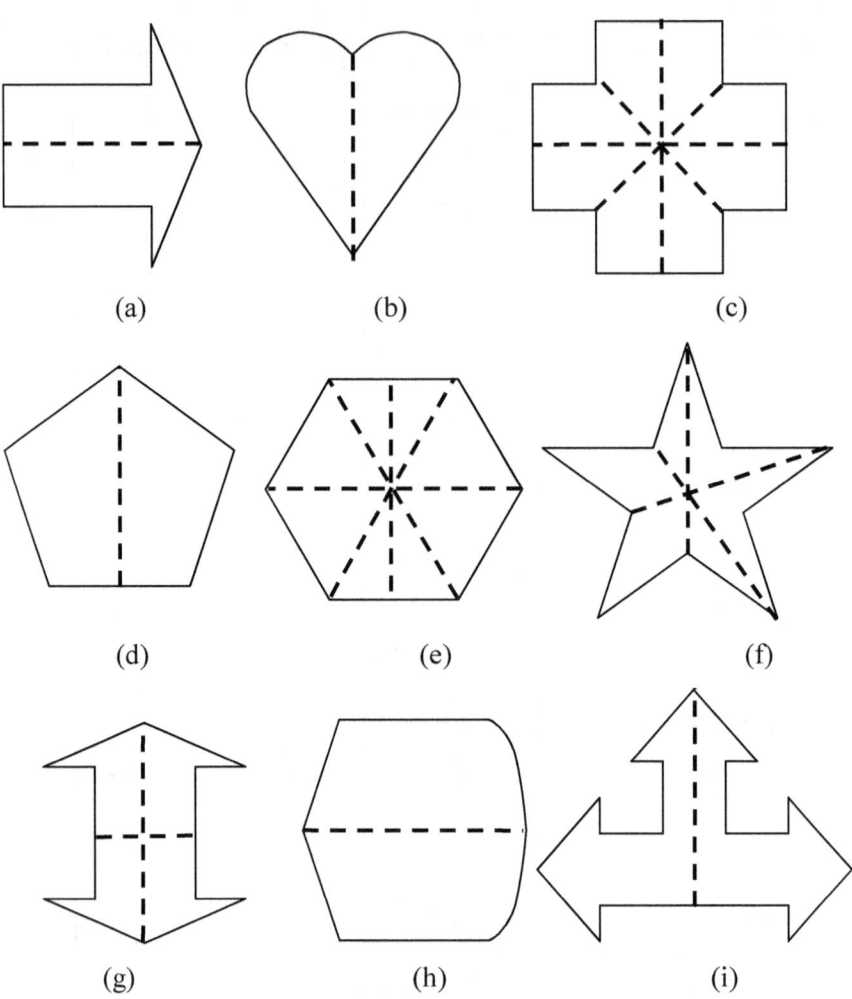

(a) (b) (c)

(d) (e) (f)

(g) (h) (i)

Exercise 12:2

1. (a) none (b) none (c) none (d) 2 (e) 4

   (f) 3 (g) 2 (h) 4 (i) infinite

2 H, I, O, X, N, S, Z.

3. A, B, C, D, E, F, G, J. K, L, M, P, Q, R, T, U, V, W, Y

4.

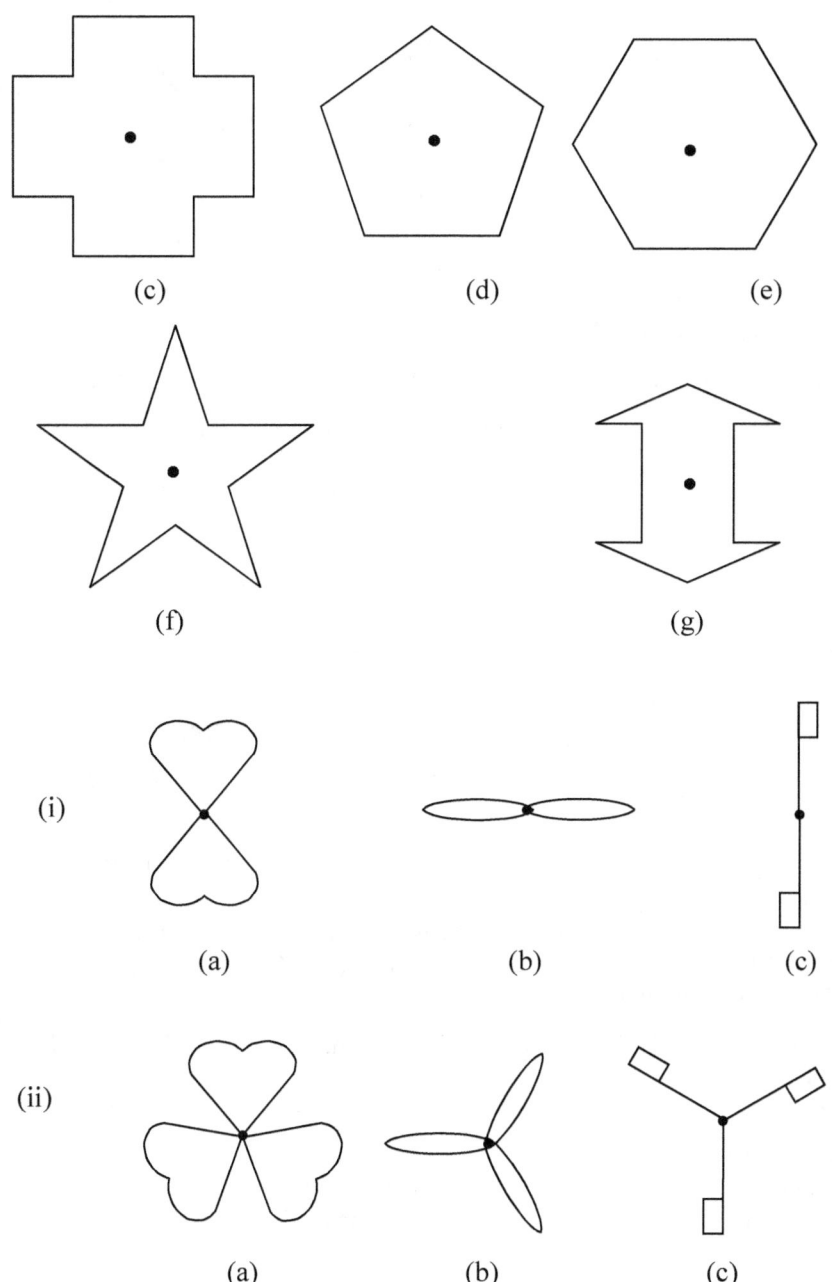

(c) (d) (e)

(f) (g)

(i)

(a) (b) (c)

(ii)

(a) (b) (c)

(iii)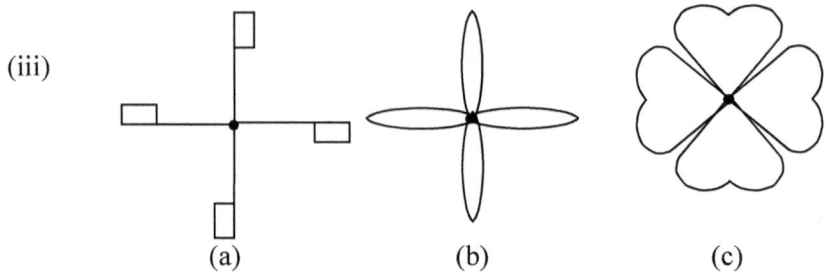

| (a) | (b) | (c) |

Skill Building Exercise 13:1

1. (a) Mah   (b) Abu   (c) Lum   (d) Abe   (e) Feh   (f) Tata

2. No. (3, 4) stand for column 3, row 4 while (4,3) stands for column 4, row 3.

Exercise 13:2

1. 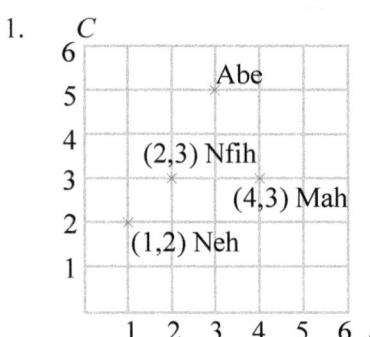   2.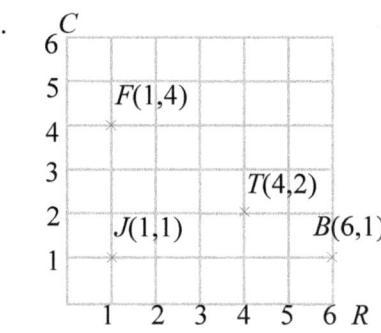

3. Bamenda (13,22), Limbe (8,8),   Mbengwi (13,20),   Mutengene (8,10),
Kumba (8,12),   Fundong (12,24),   Nkambe (18,26),   Buea (6,8), Nguti
(8,14),   Eyumojock (5,16)

4. (12,26) Wum,   (15,22) Ndop,   (10,16) Fontem,   (6,8) Buea, (17,22)
Kumbo.

Exercise 13:3

1. (a) neither   (b) vertical   (c) horizontal
   (d) neither   (e) neither   (f) vertical   2. (−7, 13)

Exercise 13:5

1. (a) Nampula   (b) Bongor   (c) Brazaville   (d) Dakar
   (e) Kampala   (f) Cape Town   (g) Conakry   (h) Tunis
2. (a) (26,25)   (b) (0,5)   (c) (12,4)   (d) (15, −5)

284

Exercise 14:1

1. (a) 6 (b) 6   2. (a) 12   (b) 12   3. (a) 8 (b) 8   4. b

Exercise 14:2

1. 8400 cm^3 2. 1440 m^3 3. 40 cm   4. 13 m
5. 10 m   6. 3000 cm^3 7. 144 m^3 8. 192

Exercise 14:3

1. 12000 cm^3 2. $10\,l$ 3. 11000 cm^3 4. 72000 cm^3
5. (a) 7000 cm^3 (b) 40000 cm^3 (c) 30 cm^3 (d) 17.4 cm^3
7. $10000\,l$ 8. $60000\,l$ 9. (a) 8 cm^3 (b) 24 cm^2

Exercise 15:1

1. 6927.2 cm^3 2.

Base radius	5 cm	10 cm
Height	20 cm	30 cm
Surface area	785.7 cm^2	2514.3 cm^2
Volume	1571.4 cm^3	9428.6 cm^3

3. 53.6 m^2 4. 1.8 cm   5. 1078 cm^3 6. 3 cm   7. 5 cm
8. 22000 cm^3 9. 3080 cm^3 10. 127.3 cm   11. 4620 cm^3 12. 220 cm^3

Exercise 15:2

1. 462 cm^2 2. 352 cm^2 3. 1584 cm^3 4. (a) 880 cm^2 (b) 1496 cm^2
5. 314 cm^3 6. 154 cm^3 7. 314.3 cm^3

Exercise 16:1

1.(a)

Data	Type of data	Method used
Name of student	qualitative	Questionnaire, interview
Class of student	quantitative	Questionnaire, interview, survey
Name of guardian	qualitative	Questionnaire, interview
Village of origin	qualitative	Questionnaire, interview
Center Number	qualitative	Questionnaire, interview
Candidate name	qualitative	Questionnaire, interview
Candidate number	qualitative	Questionnaire, interview, survey
Sex of candidate	qualitative	Census, survey
Age of candidate	quantitative	Questionnaire, interview, survey, research or archive
Religion of candidate	qualitative	Questionnaire, interview
Fee paid by candidate	quantitative	Questionnaire, interview, survey

2. (a) Questionnaire, interview (b) Questionnaire, interview (c) survey
  (d) survey   (e) census   (f) survey   (g) Questionnaire, interview
  (h) survey   (i) Questionnaire, interview (j) Questionnaire

(k) Questionnaire, interview   (l) survey
4. (i) quantitative   (ii) quantitative   (iii) qualitative   (iv) quantitative
5. (a)   185   (b)   Yes   (c)   15   (d)   2011   (e)   2016   (f)   630

Exercise 16:3

1.

Distance, x	66	67	68	69	70	71
Tally	|||	||||	++++		||||	|||		

Distance, x	66	67	68	69	70	71
Frequency, f	2	4	6	4	3	1

1.

Marks, x	Tally
49		
51		
53			
54			
56				
57				
58				
59				
60	++++
62					
63				
64				
66			

Marks, x	f
49	1
51	1
53	2
54	2
56	3
57	3
58	3
59	3
60	5
62	4
63	3
64	3
66	2

3.

Food item	Tally
r	++++ ||||
b	++++ ||||
p	++++ ++++ ||
y	++++ ++++ ++++

Food item	f
r	9
b	9
p	12
y	15

4.

weight	Tally
52	I
55	III
56	I
57	III
58	₩ III
59	III
60	₩ I
61	₩ I
62	₩ I
63	₩ II
64	₩ ₩
66	₩
69	I

weight	f
52	1
55	3
56	1
57	3
58	8
59	3
60	6
61	6
62	6
63	7
64	10
66	5
69	1

Exercise 16:10

1. (a) (i) (b) Fruit industry (c) Electricity Company (d) No
2.

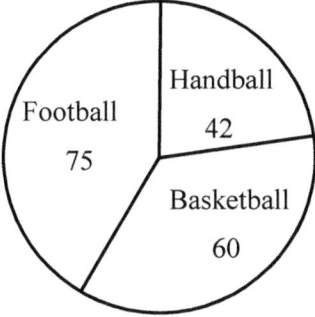

The angle for basketball is 120°

3. (a) 6000 FCFA (b) 3750 FCFA.

4.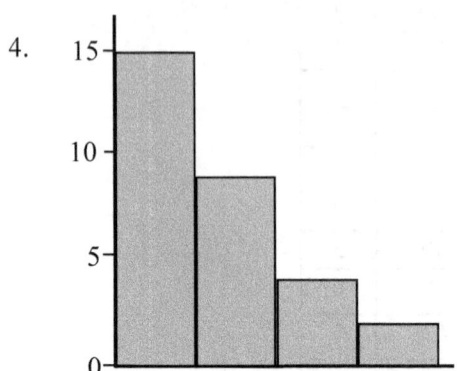

5. 41.7%

6. 14°

7.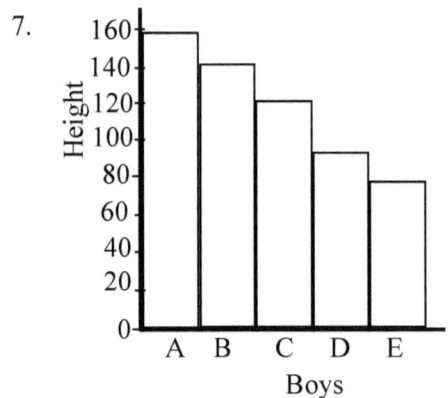

8. (a)   100°   (b)   15

9. $x = 132°, \ y = 72$

www.ingramcontent.com/pod-product-compliance
Lightning Source LLC
Chambersburg PA
CBHW071411180526
45170CB00001B/70